AF215357

Artificial Intelligence Essentials

Oliver Kramer

Artificial Intelligence
Essentials

 Springer

Oliver Kramer
Computational Intelligence Lab
Carl von Ossietzky University of Oldenburg
Oldenburg, Germany

ISBN 978-3-032-06636-7   ISBN 978-3-032-06637-4   (eBook)
https://doi.org/10.1007/978-3-032-06637-4

This Springer imprint is published by the registered company Springer Nature Switzerland AG
The registered company address is: Gewerbestrasse 11, 6330 Cham, Switzerland

If disposing of this product, please recycle the paper.

Preface

Artificial Intelligence is more important than ever before. It has become an integral part of many products and our daily lives, even replacing certain labor tasks. With the advent of powerful technologies like large language models, the field is evolving rapidly. In this book, we will explore fundamental algorithms, techniques, and key principles of AI, from machine learning and deep learning to large language models and agentic AI. A central focus is on the implementation and application of AI methods. We will not only cover the algorithmic foundations but also emphasize practical aspects, reinforcing understanding through hands-on exercises and real-world examples.

Oldenburg, Germany   Oliver Kramer
July 2025

Contents

Part II   Deep Learning

Part III   LLMs

Part IV Optimization

Chapter 1
Introduction

1.1 What Is AI?

Artificial Intelligence (AI) describes systems that learn, reason, and make decisions. Modern AI relies on deep learning and neural networks. These models handle tasks such as language processing, image recognition, and automated decision-making. Early AI used rule-based systems, but current progress comes from machine learning on large datasets. Large language models (LLMs) show recent advances. They use transformer networks to generate text, support programming, and solve reasoning tasks. Agentic AI extends these models. It connects LLMs with automation and performs multi-step tasks with minimal human input. AI systems now move from static behavior to adaptive and interactive operation. AI also operates beyond language. New methods create images, master complex games, and control robots. Research pushes AI into healthcare, finance, and scientific discovery. Its impact grows as it enters more domains. This book introduces machine learning with `scikit-learn`, `keras`, and LLMs via `ollama`. Each chapter provides practical examples and exercises for real applications.

1.2 History of AI

AI has developed through many stages. Below is a short timeline of major events:

- 1950s – Birth of AI: Alan Turing proposed machine intelligence with the Turing Test. John McCarthy introduced the term "Artificial Intelligence" at the 1956 Dartmouth Conference.

O. Kramer, *Artificial Intelligence Essentials*,
https://doi.org/10.1007/978-3-032-06637-4_1

- 1960s – Early Systems: Researchers built early programs such as ELIZA by Joseph Weizenbaum. Newell and Simon created the General Problem Solver (GPS), which showed basic problem-solving ability.
- 1970s – First AI Winter: Progress slowed because hardware was weak and expectations were unrealistic. Funding and interest declined.
- 1980s – Expert Systems: Expert systems renewed interest. These rule-based tools captured human knowledge and achieved commercial success.
- 1990s – New Breakthroughs: Faster computers and improved algorithms advanced the field. IBM's Deep Blue defeated chess champion Garry Kasparov in 1997.
- 2000s – Machine Learning Growth: Data-driven methods became dominant. Techniques such as SVMs, decision trees, and random forests proved effective. Large datasets and stronger hardware accelerated research.
- 2010s – Deep Learning: Neural networks and GPUs transformed AI. AlexNet won the 2012 ImageNet competition and triggered broad interest in deep learning. The transformer architecture by Vaswani et al. [36] reshaped NLP in 2017.
- 2020s – Large Language Models: Large models such as GPT reached new levels of language generation and understanding. They write text, produce code, and perform creative tasks. Generative AI systems such as ChatGPT demonstrated strong real-world value.

1.3   How to Use This Book

This book is designed to blend theory with practice and guide you through Hands on exploration of AI methods. Along the way, you'll find useful visual cues:

- ▓ QR codes next to Python examples link to interactive Google Colab notebooks. You'll need a Google account to run and edit them online.
- 📖 This icon indicates links to recommended reading or further literature, helping you dive deeper into the topic.
- 🖥 Each chapter concludes with a set of practical exercises to reinforce your understanding and encourage experimentation.

Info Boxes

Info Boxes provide concise yet complete insights into key concepts, methods, and actions. They are designed to enrich understanding at a glance while offering guidance that connects technical knowledge with practical and reflective thinking.

1.4 Further Reading: AI Textbooks

Below is a selection of recommended textbooks for studying various aspects of AI:

- *Introduction to Statistical Learning* by Gareth James, Daniela Witten, Trevor Hastie, and Robert Tibshirani [15]: An introduction to statistical learning.
- *The Elements of Statistical Learning* by Trevor Hastie, Robert Tibshirani, and Jerome Friedman [11]: An advanced and comprehensive reference for statistical learning theory.
- *Dive into Deep Learning* by Aston Zhang, Zachary C. Lipton, Mu Li, and Alexander J. Smola [38]: A practical, interactive guide to deep learning with code examples in `Python`.
- *Pattern Recognition and Machine Learning* by Christopher M. Bishop [4]: A comprehensive reference for pattern recognition, Bayesian methods, and machine learning theory.
- *Deep Learning* by Ian Goodfellow, Yoshua Bengio, and Aaron Courville [10]: A foundational book covering deep learning concepts, architectures, and theoretical foundations.
- *Artificial Intelligence: A Modern Approach* by Stuart Russell and Peter Norvig [28]: A widely used textbook that covers a broad range of AI topics, including search algorithms, knowledge representation, and planning.

There is a wide range of additional books and literature available for further reading.

1.5 Exercises

These exercises help you review the main ideas from this chapter.

- Questions:

    - What is AI in your own words?
    - Name important milestones in AI history.
    - Find and explore a freely available online AI textbook.

- Programming:

    - Install Python and set up a programming environment using Conda.
    - Try out Jupyter Notebook and Visual Studio Code. What are the pros and cons?
    - Open Google Colab and run a short Python script like: `print("Hello, AI!")`

Stay Informed with Articles

Read articles on AI topics from platforms like Towards Data Science to stay updated with recent developments and trends.

Part I
Machine Learning

Chapter 2
K-Nearest Neighbors

2.1 Supervised Learning

Supervised learning is a core idea in AI. A model learns to predict labels from labeled data [11]. In image classification, a model trains on many labeled images, such as cats, dogs, or cars. It learns to link visual patterns to these classes. In healthcare, supervised models estimate disease risk from structured inputs such as blood pressure, cholesterol, and age. Training uses a dataset of n labeled examples. Each input pattern $\mathbf{x}_j \in \mathbb{R}^d$ is a d-dimensional vector. The label $y_j \in \mathcal{Y}$ matches the input:

$$\mathcal{X} = \{(\mathbf{x}_1, y_1), (\mathbf{x}_2, y_2), \ldots, (\mathbf{x}_n, y_n)\}.$$

Each pattern $\mathbf{x}_j = (x_{j1}, x_{j2}, \ldots, x_{jd})$ contains d features. The value x_{ji} is the i-th feature of the j-th pattern. The aim is to learn a function $f(\mathbf{x})$ that maps inputs to outputs. During training, the model updates parameters to reduce the gap between predictions $f(\mathbf{x}_j)$ and true labels y_j. These labeled examples form the ground truth.

Each data point occupies a position in a high-dimensional space defined by its features. These features can be numerical, categorical, or mixed. A supervised model tries to learn structure in this space by finding patterns that link inputs to outputs. Such patterns can be correlational or sometimes causal. Good generalization requires understanding this structure so the model can make accurate predictions on unseen data. Supervised learning tasks fall into two groups. In classification, the model assigns each input to a discrete class, such that $f(\mathbf{x}) \in \mathcal{C}$. In regression, the model outputs a continuous value, where $f(\mathbf{x}) \in \mathbb{R}$. The success of supervised learning depends on data quality, data diversity, and the model's ability to extract stable and meaningful patterns.

© The Author(s), under exclusive license to Springer Nature Switzerland AG 2026
O. Kramer, *Artificial Intelligence Essentials*,
https://doi.org/10.1007/978-3-032-06637-4_2

2.2   KNN

The k-nearest neighbors (KNN) algorithm [7] is a simple method for classification and regression. It follows the idea that similar points tend to share similar outcomes. Given a query point $\mathbf{x}'$, the algorithm selects the k training points closest to it. The distance is often the Euclidean metric:

$$\delta(\mathbf{x}', \mathbf{x}_i) = \sqrt{\sum_{j=1}^{d} (x'_j - x_{ij})^2}$$

Here, $\mathbf{x}'$ is the input vector, $\mathbf{x}_i$ is a training point, and j indexes the features. For classification, KNN assigns the label that appears most often among the k selected neighbors.

Figure 2.1 shows the classification step. The query point receives the majority label from its three nearest neighbors. The algorithm is non-parametric because it makes no assumptions about the data distribution. This property keeps KNN simple and flexible.

Figure 2.2 presents decision boundaries for four values of k. With $k = 1$, the boundary becomes very irregular and follows individual points. With $k = 3$, the boundary smooths out and reacts less to noise. Larger values such as 50 and 200 produce broad and stable regions, which reduces overfitting and moves the model toward underfitting.

Manhattan Distance
The Manhattan distance sums absolute differences between features:

$$\delta(\mathbf{x}', \mathbf{x}_i) = \sum_{j=1}^{d} |x'_j - x_{ij}|$$

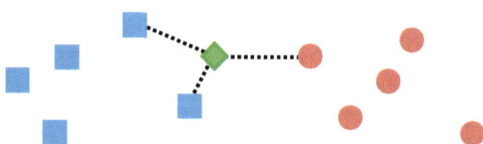

Fig. 2.1 Illustration of KNN classification: the new pattern (green diamond) is classified as blue square because, among its $k = 3$ nearest neighbors, two of three are labeled blue square, one as red circle

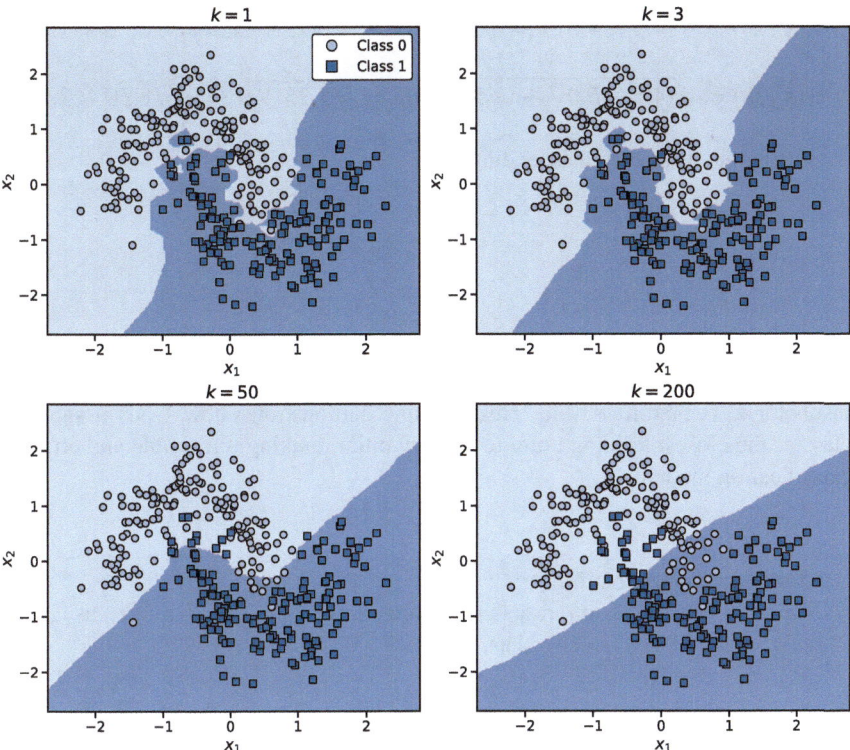

Fig. 2.2 Decision boundaries of a k-NN classifier on the moons dataset for $k = 1, 3, 50$, and 200. Darker regions correspond to the circle class and lighter regions to the square class. Small values of k lead to highly flexible, local decision boundaries, whereas larger values increasingly smooth the classifier and emphasize global structure

2.3 Hands on KNN

Consider a dataset with three labeled training points:

Point	x_1	x_2	Class
$\mathbf{x}_1$	1.0	2.0	Blue
$\mathbf{x}_2$	4.0	1.0	Blue
$\mathbf{x}_3$	6.0	5.0	Red

We classify a new point $\mathbf{x}' = (4.5, 2.5)$ using $k = 1$-NN. The Euclidean distances to each training point are calculated as:

$$\delta(\mathbf{x}', \mathbf{x}_1) = \sqrt{(4.5 - 1.0)^2 + (2.5 - 2.0)^2} = \sqrt{12.25 + 0.25} = \sqrt{12.50} \approx 3.54$$

$$\delta(\mathbf{x}', \mathbf{x}_2) = \sqrt{(4.5 - 4.0)^2 + (2.5 - 1.0)^2} = \sqrt{0.25 + 2.25} = \sqrt{2.50} \approx 1.58$$

$$\delta(\mathbf{x}', \mathbf{x}_3) = \sqrt{(4.5 - 6.0)^2 + (2.5 - 5.0)^2} = \sqrt{2.25 + 6.25} = \sqrt{8.50} \approx 2.92$$

Sorting the distances in ascending order:

Point	$\delta(\mathbf{x}', \mathbf{x}_i)$
$\mathbf{x}_2$	1.58
$\mathbf{x}_3$	2.92
$\mathbf{x}_1$	3.54

Since we use $k = 1$, the nearest neighbor is point $\mathbf{x}_2$ (Blue). The classification result for $\mathbf{x}'$ is therefore Blue. This example demonstrates how 1-NN assigns the class of the closest training point to a query point, making it a simple and effective classification method.

Min-Max Normalization
Min-max normalization rescales feature values to a fixed range, usually between 0 and 1. It is defined as:

$$\hat{x} = \frac{x - \min(X)}{\max(X) - \min(X)}$$

Here, x is a single value from the feature vector $X = \{x_1, x_2, \ldots, x_n\}$. The terms $\min(X)$ and $\max(X)$ are the minimum and maximum values of that feature in the dataset. This method keeps relative distances intact. It is important in KNN because feature scales influence distance computations.

2.4   Precision and Recall

A classification model is often evaluated with metrics derived from the confusion matrix. This matrix lists prediction outcomes in four groups: true positives (TP), false positives (FP), false negatives (FN), and true negatives (TN). The structure is:

	Predicted Positive	Predicted Negative
Actual Positive	TP	FN
Actual Negative	FP	TN

Accuracy measures the share of correct predictions across both classes:

$$\text{Accuracy} = \frac{TP + TN}{TP + TN + FP + FN}$$

Accuracy can be misleading when the dataset is imbalanced. Precision measures how many predicted positive cases are correct:

$$\text{Precision} = \frac{TP}{TP + FP}$$

Recall, or sensitivity, measures how many actual positive cases the model identifies:

$$\text{Recall} = \frac{TP}{TP + FN}$$

Consider a model tested on 100 samples. Assume 40 true positives, 5 false positives, 10 false negatives, and 45 true negatives. The accuracy is $\frac{40+45}{100} = 0.85$. Precision is $\frac{40}{45} \approx 0.89$. Recall is $\frac{40}{50} = 0.8$.

Overfitting Versus Underfitting
Underfitting occurs when a model is too simple to capture data structure. It performs poorly on both training and test sets. Overfitting occurs when a model is too complex and memorizes noise. It performs poorly on unseen data. In KNN, a very small k (e.g., $k = 1$) often causes overfitting. The model reacts strongly to noise and outliers. A very large k (e.g., $k = 200$) may cause underfitting because the model smooths over local patterns and ignores fine class boundaries.

2.5 Cross-Validation

Cross-validation and grid search are key methods for model evaluation and hyperparameter tuning. Together, they improve generalization and help prevent overfitting. Cross-validation estimates performance on unseen data. In k-fold cross-validation, shown in Fig. 2.3, the dataset is split into k equal folds. The model trains on $k - 1$ folds and validates on the remaining fold. This repeats k times so each fold serves once as validation. The final score is the average across all runs. This produces a more stable estimate than a single train-test split. Variants include stratified k-fold, which keeps class ratios stable, and leave-one-out cross-validation (LOOCV), where each fold contains one sample. Cross-validation is useful for small datasets and helps detect overfitting through repeated testing on unseen data.

Grid search tunes hyperparameters by testing all combinations of predefined values. Each combination is evaluated with cross-validation. For KNN, grid search can

Fig. 2.3 Illustration of
3-fold cross-validation:
Across three CV rounds,
each fold is selected once as
the validation set and twice
as part of the training set

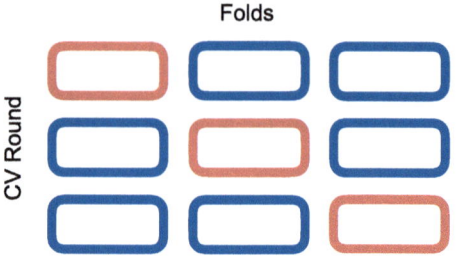

select the best number of neighbors by testing values such as $k \in \{1, 3, 5, 7, 9\}$. The
value with the best average accuracy is chosen. Grid search is exhaustive but becomes
expensive when many parameters or wide ranges are used.

KNN Regression
KNN regression predicts continuous values by averaging the outputs of the k
nearest points. Given a query $\mathbf{x}' \in \mathbb{R}^d$, the model finds the k closest neighbors
and computes:

$$f(\mathbf{x}') = \frac{1}{k} \sum_{j=1}^{k} y_j$$

where y_j is the target value of the j-th neighbor. This non-parametric method
assumes that nearby points have similar outputs. It works well for non-linear
regression.

Performance is evaluated with the Mean Squared Error (MSE):

$$\text{MSE} = \frac{1}{n} \sum_{j=1}^{n} (y_j - f(\mathbf{x}_j))^2$$

Here, n is the number of test samples, y_j is the true value, and $f(\mathbf{x}_j)$ is the
prediction for the j-th sample.

2.6   KNN in `scikit-learn`

The following script in `scikit-learn` [23] demonstrates how to train a KNN
classifier on the Wine dataset, evaluate it using classification metrics, and tune the
hyperparameter k using grid search with cross-validation.

```
from sklearn.datasets import load_wine
from sklearn.model_selection import train_test_split,
    GridSearchCV
from sklearn.neighbors import KNeighborsClassifier
from sklearn.metrics import classification_report,
    confusion_matrix, ConfusionMatrixDisplay

# Load Wine dataset and split into train/test sets
wine = load_wine()
X, y = wine.data, wine.target
X_train, X_test, y_train, y_test = train_test_split(
    X, y, test_size=0.3, random_state=42)

# Grid search over different k values
param_grid = {'n_neighbors': [1, 3, 5, 7, 9]}
grid_search = GridSearchCV(KNeighborsClassifier(), param_grid,
    cv=5)
grid_search.fit(X_train, y_train)

# Use the best model to predict test data
best_knn = grid_search.best_estimator_
y_pred = best_knn.predict(X_test)

# Evaluate the model
print("Best k:", grid_search.best_params_['n_neighbors'])
print("Classification Report:")
print(classification_report(y_test, y_pred,
    target_names=wine.target_names))

# Show confusion matrix
cm = confusion_matrix(y_test, y_pred)
ConfusionMatrixDisplay(confusion_matrix=cm,
                       display_labels=wine.target_names).plot()
```

This script executes the full KNN pipeline. It trains the model, tunes hyperparameters with cross-validation, evaluates performance, and creates visualizations. All steps run on the Wine dataset. The random seed is set to 42 to ensure reproducible data splits and model behavior. The number 42 is often used as a humorous reference to Douglas Adams' novel *The Hitchhiker's Guide to the Galaxy*, where it is described as "the answer to the ultimate question of life, the universe, and everything" [1].

2.7   Exercises

The following exercises help you understand KNN, normalization, model evaluation, and how to improve a classifier.

- Questions:
    - What does the KNN algorithm do? How does the number k affect the result?
    - Why do we normalize features before using KNN?
    - What is the difference between overfitting and underfitting?

- What are precision and recall? When are they important?
- Why do we use cross-validation when training a model?

• Hands on:

- Look at this dataset and classify the point (3.5, 2.5) using 3-nearest neighbors and Euclidean distance:

Point	x_1	x_2	Label
x_1	1.0	3.0	Blue
x_2	3.0	2.0	Red
x_3	5.0	1.0	Red
x_4	4.0	4.0	Blue

Calculate the distances and decide which class wins.

• Programming:

- Write your own version of KNN in Python. Test it on the Iris dataset.
- Try different values for k and see how it changes the results on the Breast Cancer dataset.
- Use GridSearchCV to find the best k and distance type.
- Show the confusion matrix for your best model and explain what it means.

Geoffrey Hinton - An AI Pioneer
Explore the life and work of Geoffrey Hinton, one of the key figures in the development of AI. Investigate his groundbreaking contributions to neural networks and follow his academic and research career across institutions and beyond.

Chapter 3
K-Means

3.1 Unsupervised Learning

Unsupervised learning analyzes data that contain only input features $X = \{\mathbf{x}_1, \ldots, \mathbf{x}_n\}$. No labels are provided. The goal is to detect hidden patterns or structures. Clustering methods, such as k-means, group points into clusters based on feature similarity. Points in the same cluster are more similar to each other than to points in other clusters. Dimensionality reduction techniques map high-dimensional data into a lower-dimensional space while keeping essential structure. Because no predefined outputs exist, unsupervised learning is useful for exploratory analysis, anomaly detection, and feature extraction.

3.2 K-Means

The k-means algorithm [21] is an unsupervised method that partitions data into k clusters. The dataset contains points $\mathbf{x}_1, \mathbf{x}_2, \ldots, \mathbf{x}_n$ with $\mathbf{x}_j \in \mathbb{R}^d$. The algorithm proceeds in an iterative loop:

1. Initialize k centroids $\mathbf{c}_1, \ldots, \mathbf{c}_k$ at random.
2. Assign each point $\mathbf{x}_j$ to the nearest centroid:

$$l = \arg\min_{m} \|\mathbf{x}_j - \mathbf{c}_m\|^2$$

3. Update each centroid $\mathbf{c}_l$ by taking the mean of all points in cluster l:

$$\mathbf{c}_l = \frac{1}{|C_l|} \sum_{\mathbf{x}_j \in C_l} \mathbf{x}_j$$

where C_l is the set of points assigned to l.

© The Author(s), under exclusive license to Springer Nature Switzerland AG 2026
O. Kramer, *Artificial Intelligence Essentials*,
https://doi.org/10.1007/978-3-032-06637-4_3

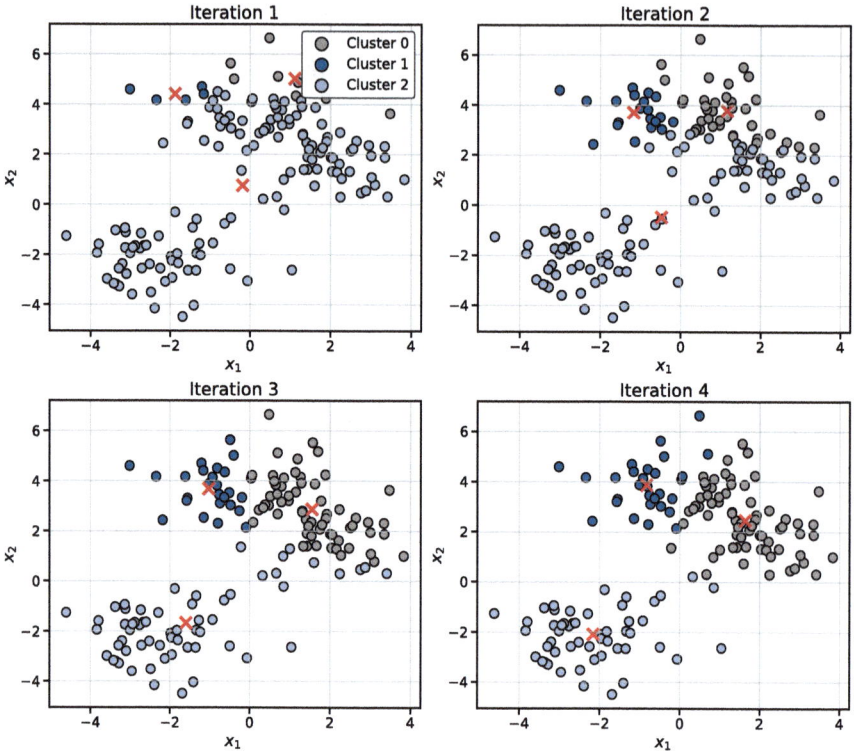

Fig. 3.1 K-means clustering with $k = 3$ on a synthetic dataset with two features x_1 and x_2. Data points are colored according to their current cluster assignment, while red crosses denote the centroids. The figure illustrates successive iterations of the algorithm as the cluster structure stabilizes

4. Repeat steps 2 and 3 until assignments stop changing.

Figure 3.1 shows four iterations of k-means on a synthetic dataset. The centroids, marked with red crosses, shift as points move between clusters. The process stabilizes when no point changes its assignment.

Intra-Cluster Variance

Intra-cluster variance measures how closely points in a cluster lie around the centroid. Low variance indicates compact clusters. It is defined as:

$$\text{Intra-cluster variance} = \sum_{l=1}^{k} \frac{1}{|C_l|} \sum_{\mathbf{x}_j \in C_l} \|\mathbf{x}_j - \mathbf{c}_l\|^2$$

Here, C_l is the set of points in cluster l, c_l is the centroid, and $x_j \in \mathbb{R}^d$ are the points. Minimizing this value produces cohesive clusters. Inter-cluster variance should be high to keep centroids well separated and clusters clearly distinct.

3.3 Hands on K-Means

Consider a dataset with four data points in two-dimensional space:

Point	x_1	x_2
x_1	1.0	2.0
x_2	3.0	1.0
x_3	4.0	4.0
x_4	5.0	2.0

We aim to partition these points into two clusters ($k = 2$). The algorithm starts by initializing two random centroids. Assume the initial centroids are:

$$c_1 = (1.0, 2.0), \quad c_2 = (5.0, 2.0)$$

First, assign Points to Nearest Centroid. Using the Euclidean distance, each point is assigned to the nearest centroid:

$$\delta(x_1, c_1) = \sqrt{(1.0 - 1.0)^2 + (2.0 - 2.0)^2} = 0.0$$

$$\delta(x_1, c_2) = \sqrt{(1.0 - 5.0)^2 + (2.0 - 2.0)^2} = 4.0$$

Since $\delta(x_1, c_1) < \delta(x_1, c_2)$, point x_1 is assigned to cluster 1. Repeating this for the remaining points, we obtain the assignments:

$$\text{Cluster 1: } \{x_1, x_2\}, \quad \text{Cluster 2: } \{x_3, x_4\}$$

Second, the centroids are updated. New centroids are computed as the mean of the assigned points:

$$c_1 = \left(\frac{1.0 + 3.0}{2}, \frac{2.0 + 1.0}{2} \right) = (2.0, 1.5)$$

$$c_2 = \left(\frac{4.0 + 5.0}{2}, \frac{4.0 + 2.0}{2} \right) = (4.5, 3.0)$$

Steps 1 and 2 are repeated until centroids no longer change. The final clusters separate the points into two distinct groups. This example illustrates the iterative process of k-means, where points are reassigned and centroids are updated until convergence.

3.4   K-Means in `Python`

Below is a short implementation in `Python`.

```python
def k_means(X, k, iterations=4):
    # Randomly initialize k centroids within the feature range
        of X
    centroids = np.random.uniform(low=X.min(axis=0),
        high=X.max(axis=0), size=(k, X.shape[1]))
    all_centroids = [centroids.copy()]  # Store initial
        centroids
    all_clusters = []  # To store cluster assignments at each
        iteration

    for _ in range(iterations):
        # Assign each point to the nearest centroid (Euclidean
            distance)
        clusters = [np.argmin([np.linalg.norm(x - c) for c in
            centroids]) for x in X]

        # Compute new centroids as the mean of points in each
            cluster
        new_centroids = [X[np.array(clusters) ==
            i].mean(axis=0) for i in range(k)]
        centroids = np.array(new_centroids)  # Update centroids
        all_centroids.append(centroids.copy())  # Track
            centroids over time
        all_clusters.append(clusters)  # Track cluster
            assignments

    return all_clusters, all_centroids  # Return history of
        assignments and centroids

# Run K-Means
k = 3
clusters_over_time, centroids_over_time = k_means(X, k)
```

The function k_means() takes a dataset X, the number of clusters k, and a fixed number of iterations. Unlike the classic variant, which initializes centroids by randomly selecting k data points from the dataset and runs until convergence, this version initializes centroids uniformly within the data range and performs a fixed number of iterations. In each iteration, the algorithm assigns every data point to the nearest centroid using the Euclidean distance and updates each centroid as the mean of all points currently assigned to its cluster.

3.5 Exercises

The following exercises help you understand how k-means clustering works, how to choose parameters, and how to evaluate results.

- Questions:

  - Why do we need to choose the number of clusters k before running k-means? How can we find a good value?
  - What problems can k-means have with strange-shaped or overlapping clusters?
  - Why does the starting position of the cluster centers matter?

- Hands on:

  - Use the following four points in 2D space:

Point	x_1	x_2
x_1	2.0	1.0
x_2	3.0	3.0
x_3	6.0	2.0
x_4	7.0	4.0

  Start with cluster centers at (2.0, 1.0) and (6.0, 2.0). Assign each point to the closest center, then update the centers using the mean of each group.

- Programming:

  - Write your own k-means algorithm in Python and test it on a dataset from `scikit-learn`.
  - Try different values for k and plot the clusters to see how the result changes.
  - Look up what `MiniBatchKMeans` does in `scikit-learn` and when it is useful.

AI in Movies
Discover which films have incorporated AI in their creation, whether in plot development, dialogue generation, or visual effects, and explore how these innovations are shaping the future of the movie industry.

Part II
Deep Learning

Chapter 4
Multi-layer Perceptrons

4.1 Simple Perceptrons

Neural networks form the basis of modern deep learning. They draw inspiration from the brain and appear in tasks such as image recognition, language processing, autonomous driving, and robotics. The perceptron [26] is the simplest neural model and was introduced in the 1950s.

Natural Neurons

Neurons are basic units of the brain. Each neuron has dendrites for inputs, a soma as the cell body, and an axon for output. Information travels through synapses, which can strengthen or weaken. Hebbian learning captures this idea and is often stated as *cells that fire together wire together*. When both connected neurons activate at the same time, the synapse becomes stronger. This plasticity supports learning, memory, and the formation of complex neural circuits.

A perceptron takes input values, multiplies them by weights, adds a bias, and applies an activation function. This can be summarized as:

```
Input times weight, add bias, activate.
```

Using the ReLU activation, the computation becomes:

$$y = \text{ReLU}(w \cdot x + b) \tag{4.1}$$

Here, $x \in \mathbb{R}^d$ is the input vector, $w \in \mathbb{R}^d$ the weight vector, b the bias, and y the output. ReLU is defined as:

$$\text{ReLU}(z) = \max(0, z)$$

It passes positive values and blocks negative ones. This behavior helps reduce the vanishing gradient problem. The original perceptron used a step function and acted as a binary classifier with a linear boundary. This restricts it to linearly separable tasks. Even with this limit, the idea of updating weights based on errors was crucial. It laid the groundwork for backpropagation, which trains modern multi-layer networks.

Perceptron Limitation

A single-layer perceptron can model linearly separable Boolean functions such as AND, OR, and NAND. It cannot model XOR because no straight line separates its classes. This shortcoming motivated the move to multi-layer perceptrons, which solve such tasks through hidden layers and non-linear feature combinations.

4.2   MLPs

Multi-Layer Perceptrons (MLPs) extend the perceptron by stacking dense layers, see Fig. 4.2. Each dense layer computes:

$$\mathbf{y} = \sigma(\mathbf{W}\mathbf{x} + \mathbf{b})$$

Here, $\mathbf{W}$ is the weight matrix, $\mathbf{x}$ the input vector, $\mathbf{b}$ the bias, and σ a non-linear activation. By chaining several such layers, MLPs model complex non-linear relationships. Activation functions add non-linearity and allow the network to approximate complex functions. Common choices include ReLU, tanh, and sigmoid, see Fig. 4.1. The sigmoid function is:

Fig. 4.1 Visualization of the sigmoid, tanh, and ReLU activation functions. The sigmoid curve (blue) smoothly transitions from 0 to 1, the tanh curve (orange) spans values from -1 to 1, and the ReLU function (green) is zero for negative inputs and increases linearly for positive inputs

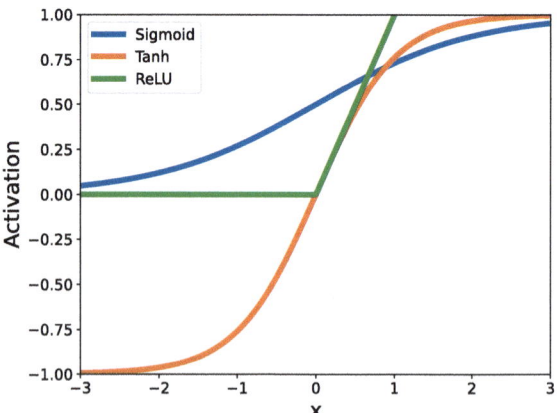

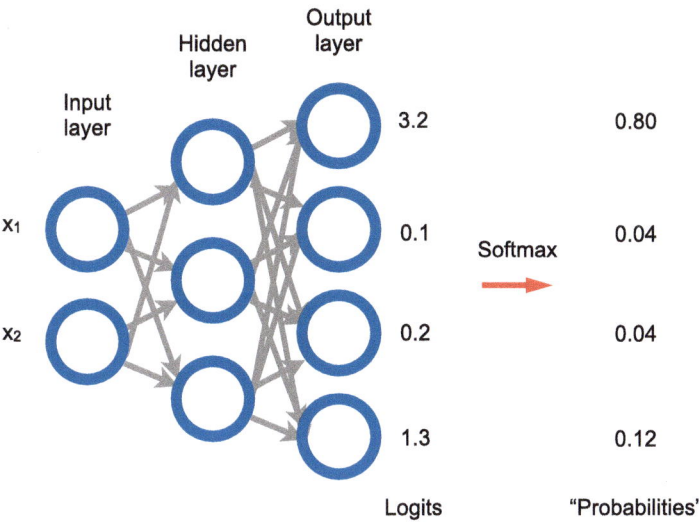

Fig. 4.2 A Multi-Layer Perceptron (MLP) with input, hidden, and output layers. Dense layers transform inputs through learned weights and non-linear activations. The final dense layer produces logits, which are passed through a Softmax function to obtain class probabilities

$$\sigma(x) = \frac{1}{1 + e^{-x}}$$

In classification, the output layer often uses Softmax. It converts logits into a probability distribution:

$$\text{Softmax}(z_i) = \frac{e^{z_i}}{\sum_{j=1}^{K} e^{z_j}}, \quad i = 1, \ldots, K$$

Softmax ensures non-negative outputs that sum to 1. Consider a small dense layer with 2 inputs and 3 outputs:

$$\mathbf{x} = \begin{bmatrix} 1 \\ 2 \end{bmatrix}, \quad \mathbf{W} = \begin{bmatrix} 0.5 & -1.0 \\ 1.5 & 0.0 \\ -0.5 & 2.0 \end{bmatrix}, \quad \mathbf{b} = \begin{bmatrix} 0.0 \\ 0.5 \\ -1.0 \end{bmatrix}$$

The weighted sum becomes:

$$\mathbf{z} = \mathbf{W}\mathbf{x} + \mathbf{b} = \begin{bmatrix} -1.5 \\ 2.0 \\ 2.5 \end{bmatrix} \implies \mathbf{y} = \text{ReLU}(\mathbf{z}) = \begin{bmatrix} 0 \\ 2.0 \\ 2.5 \end{bmatrix}$$

An MLP is a feedforward network with stacked dense layers. Non-linear activations in hidden layers allow it to model complex relations and solve tasks beyond the reach of a single perceptron.

One-Hot Encoding

One-hot encoding converts each categorical label into a binary vector. For K classes, each label becomes a vector of length K with a single 1 at the class index. For example:

$$\text{gripper} \rightarrow [1, 0, 0, 0]$$
$$\text{lidar} \rightarrow [0, 1, 0, 0]$$
$$\text{motor} \rightarrow [0, 0, 1, 0]$$
$$\text{camera} \rightarrow [0, 0, 0, 1]$$

This format is used in networks with Softmax output and categorical cross-entropy loss. It enables the model to compute class probabilities and match them to the correct class.

4.3   Backpropagation

Backpropagation [27] is the standard training algorithm for neural networks. It computes how each weight contributes to the total error by propagating gradients through the network. Training proceeds in two stages. In the forward pass, the model computes predictions. In the backward pass, it computes gradients of the loss and updates the weights. For classification, the logits are passed through Softmax. Predictions are compared to the true labels using cross-entropy loss. For a class label $y \in \{1, \ldots, K\}$ and predicted probabilities $\hat{\mathbf{y}} \in \mathbb{R}^K$, the loss is:

$$L = -\log \hat{y}_y$$

Weights are updated with gradient descent:

$$\mathbf{w} \leftarrow \mathbf{w} - \eta \cdot \nabla L(\mathbf{w})$$

Here, η is the learning rate, and $\nabla L(\mathbf{w})$ is the gradient of the loss with respect to the weights. Backpropagation adjusts model parameters over many epochs. Each epoch performs updates over mini-batches. This gradual process reduces loss and improves accuracy.

Stochastic Gradient Descent
Stochastic Gradient Descent (SGD) [18] updates parameters using a single training example. This produces noisy updates and a fluctuating loss. However, the noise can help escape shallow minima and saddle points. Mini-batch gradient descent, using small batches such as 32 or 128 samples, is widely used. It reduces variance compared to pure SGD and improves convergence speed and stability.

4.4 Hands on Backpropagation

Consider a simple perceptron with one input, one output, a ReLU activation function, and no bias. We use the Mean Squared Error (MSE) as the loss function. Let the input be $x = 2$, the target output $y = 4$, the initial weight $w = 1.0$, and the learning rate $\eta = 0.1$. In the forward pass, compute the predicted output:

$$z = w \cdot x = 1.0 \cdot 2 = 2$$

$$y_{\text{pred}} = \text{ReLU}(z) = \max(0, z) = 2$$

Compute the loss using the MSE:

$$L = \frac{1}{2}(y_{\text{pred}} - y)^2 = \frac{1}{2}(2 - 4)^2 = 2$$

The factor $\frac{1}{2}$ is included to simplify the derivative of the squared error, since the derivative of $\frac{1}{2}(y_{\text{pred}} - y)^2$ with respect to y_{pred} is simply $y_{\text{pred}} - y$, avoiding an extra factor of 2. In the backward pass, apply the chain rule to compute the gradient of the loss with respect to the weight:

$$\frac{\partial L}{\partial w} = \underbrace{\frac{\partial L}{\partial y_{\text{pred}}}}_{\text{Loss function}} \cdot \underbrace{\frac{\partial y_{\text{pred}}}{\partial z}}_{\text{Activation function}} \cdot \underbrace{\frac{\partial z}{\partial w}}_{\text{Linear term}}$$

$$\frac{\partial L}{\partial y_{\text{pred}}} = y_{\text{pred}} - y = 2 - 4 = -2$$

$$\frac{\partial y_{\text{pred}}}{\partial z} = \text{ReLU}'(z) = 1 \quad (\text{since } z = 2 > 0)$$

$$\frac{\partial z}{\partial w} = x = 2$$

$$\Rightarrow \frac{\partial L}{\partial w} = (-2) \cdot 1 \cdot 2 = -4$$

Update the weight using gradient descent:

$$w_{\text{new}} = w - \eta \cdot \frac{\partial L}{\partial w} = 1.0 - 0.1 \cdot (-4) = 1.0 + 0.4 = 1.4$$

Compute the updated prediction:

$$z_{\text{new}} = 1.4 \cdot 2 = 2.8, \quad y_{\text{new}} = \text{ReLU}(z_{\text{new}}) = 2.8$$

The new prediction $y_{\text{new}} = 2.8$ is closer to the target $y = 4$, demonstrating how backpropagation enables learning by adjusting weights to reduce error.

Dropout
Dropout [32] is a regularization method that reduces overfitting in neural networks. During training, it randomly removes a fraction of neurons in a layer by setting their output to zero. The removal follows a Bernoulli distribution. Each neuron is kept with probability p and dropped with probability $1 - p$. This forces the network to learn redundant representations and improves generalization. At test time, all neurons remain active, and their outputs are scaled to match training behavior.

4.5   MLP in `scikit-learn`

The `digits` dataset of handwritten digits (1797 samples of size 8×8) can also be classified using `scikit-learn`'s `MLPClassifier`. Here we fetch the data, normalize and split it, then train a two-hidden-layer perceptron with SGD:

```
from sklearn.datasets import load_digits
from sklearn.model_selection import train_test_split
from sklearn.preprocessing import StandardScaler
from sklearn.neural_network import MLPClassifier
from sklearn.metrics import accuracy_score,
    classification_report

# Load digits dataset (64 features, 8x8 images)
digits = load_digits()
X, y = digits.data, digits.target

# Normalize inputs to [0, 1]
X = X.astype("float32") / 16.0

# Train/test split
```

```
X_train, X_test, y_train, y_test = train_test_split(
    X, y, test_size=0.2, random_state=42
)

# Standardize
scaler = StandardScaler()
X_train = scaler.fit_transform(X_train)
X_test = scaler.transform(X_test)

# MLP model
mlp = MLPClassifier(
    hidden_layer_sizes=(64, 32),
    activation='relu',
    solver='sgd',
    learning_rate_init=0.01,
    batch_size=32,
    max_iter=50,
    random_state=42,
    verbose=True
)
mlp.fit(X_train, y_train)
```

This script builds an MLP with two hidden layers (64 and 32 units) and ReLU activations, optimizes with stochastic gradient descent, and reports accuracy and a full classification report on the held-out test set. Preprocessing includes scaling each feature to zero mean and unit variance.

4.6 Exercises

The following exercises help you understand the perceptron, backpropagation, and how neural networks are trained.

- Questions:

  - What is a simple perceptron? What parts does it have?
  - Why can't a perceptron learn the XOR function? How can this be fixed?
  - What is backpropagation and why is it important for training?
  - Why do we use activation functions in neural networks?

- Hands on:

  - Suppose we have a perceptron with a sigmoid activation. The input is $x = 10$, the target output is $y = 0.95$, the initial weight is $w = 0.05$, and the learning rate is $\eta = 0.2$.
  - Do one training step: calculate the output, compute the squared error, find the gradient, update the weight using gradient descent, and calculate the new output.

- Programming:

  - Use `scikit-learn`'s `MLPClassifier` to classify the Digits dataset. Try changing the number of hidden layers, the activation function, and the learning rate.
  - Try the same idea using `Keras` with the MNIST dataset.
  - Plot training and validation accuracy to see how the model performs with different settings.

Watch Keynotes and Talks

Watch AI-related YouTube videos, such as keynotes and TED talks from industry leaders like the CEO of NVIDIA. Check recent events online to find relevant talks.

Chapter 5
Convolutional Neural Networks

5.1 Image Recognition

Convolutional Neural Networks (CNNs) [19] transformed image recognition by learning patterns such as edges, textures, and shapes directly from raw pixels. Their layered structure, which combines convolution and pooling, enables efficient feature extraction and scales well to complex image data. CNNs remove the need for handcrafted features and learn from raw inputs in a data-driven way. This shift led to major gains in accuracy and robustness across visual recognition tasks. A major breakthrough came with AlexNet [17],[1] which showed the power of deep CNNs on the ImageNet dataset. Since then, CNNs have become standard tools in healthcare, autonomous driving, and security.

5.2 Convolutional Layers

Convolutional layers are the core of CNNs. They extract local features from input data, such as images. Each layer applies a set of filters, also called kernels, that slide across the input. These filters are small matrices, often 3×3 or 5×5, and process the input region by region. Figure 5.1 shows the process: the filter moves over the image, covers local regions called receptive fields, multiplies its values with the covered pixels, and sums the result. This produces one value in the output feature map.

---

[1] A. Krizhevsky, I. Sutskever, and G. E. Hinton, ImageNet Classification with Deep Convolutional Neural Networks, Advances in Neural Information Processing Systems, vol. 25, 2012. Available at: https://papers.nips.cc/paper_files/paper/2012/file/c399862d3b9d6b76c8436e924a68c45b-Paper.pdf.

© The Author(s), under exclusive license to Springer Nature Switzerland AG 2026   31
O. Kramer, *Artificial Intelligence Essentials*,
https://doi.org/10.1007/978-3-032-06637-4_5

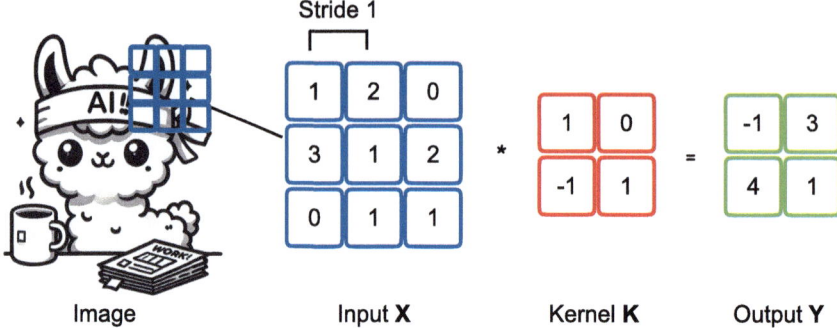

Fig. 5.1 Illustration of the convolution operation with stride 1. A kernel K is slid over the input X, and at each position an element-wise multiplication followed by summation produces one entry of the output Y

For a single-channel input and a single filter **K**, the convolution is:

$$y(i, j) = \sum_{m,n} \mathbf{K}(m, n) \cdot \mathbf{X}(i + m, j + n)$$

Here, **X** is the input map, **K** the 2D filter, and $y(i, j)$ the output at position (i, j). The summation runs over the spatial size of the filter. This is a special case of multi-channel convolution where both inputs and filters have one channel.

This repeated operation extracts patterns and allows the network to detect edges, textures, and shapes. Lower layers detect simple structures, while deeper layers detect more complex ones. The step size of the filter is called stride. It has a vertical and a horizontal component. A stride greater than 1 reduces computation and memory use. To avoid shrinking output dimensions, the network can pad the input with zeros. This is called zero padding. For example, padding with $P = 1$ adds a one-pixel border of zeros. This helps preserve spatial resolution when using small filters like 3×3.

Filter Banks

A filter bank is a set of filters that scan the input to extract different features. Each filter uses all input channels (such as RGB channels or feature maps) and produces one output map. For example, with 128 input channels and 256 filters, each filter combines all 128 channels to create one output map, producing 256 feature maps in total. The filters learn patterns such as edges or textures. The full set of feature maps captures many aspects of the input and is passed to the next layer for deeper processing.

5.3 Hands on Convolution

Consider a single-channel 3×3 input image and a 2×2 filter. We use a stride of 1 and no padding. Let the input be:

$$\mathbf{X} = \begin{bmatrix} 1 & 2 & 0 \\ 3 & 1 & 2 \\ 0 & 1 & 1 \end{bmatrix}$$

and the filter:

$$\mathbf{K} = \begin{bmatrix} 1 & 0 \\ -1 & 1 \end{bmatrix}$$

We perform the convolution by sliding the filter over the input matrix. At each location, we compute the element-wise product and sum. Step 1 (top-left region of $\mathbf{X}$):

$$\begin{bmatrix} 1 & 2 \\ 3 & 1 \end{bmatrix} * \begin{bmatrix} 1 & 0 \\ -1 & 1 \end{bmatrix} = (1 \cdot 1) + (2 \cdot 0) + (3 \cdot -1) + (1 \cdot 1) = -1$$

Step 2 (top-middle region):

$$\begin{bmatrix} 2 & 0 \\ 1 & 2 \end{bmatrix} * \begin{bmatrix} 1 & 0 \\ -1 & 1 \end{bmatrix} = (2 \cdot 1) + (0 \cdot 0) + (1 \cdot -1) + (2 \cdot 1) = 3$$

Step 3 (bottom-left region):

$$\begin{bmatrix} 3 & 1 \\ 0 & 1 \end{bmatrix} * \begin{bmatrix} 1 & 0 \\ -1 & 1 \end{bmatrix} = (3 \cdot 1) + (1 \cdot 0) + (0 \cdot -1) + (1 \cdot 1) = 4$$

Step 4 (bottom-middle region):

$$\begin{bmatrix} 1 & 2 \\ 1 & 1 \end{bmatrix} * \begin{bmatrix} 1 & 0 \\ -1 & 1 \end{bmatrix} = (1 \cdot 1) + (2 \cdot 0) + (1 \cdot -1) + (1 \cdot 1) = 1$$

The output feature map is:

$$\mathbf{Y} = \begin{bmatrix} -1 & 3 \\ 4 & 1 \end{bmatrix}$$

This demonstrates how a CNN filter extracts features from an input by scanning small regions and applying learned weights.

Data Augmentation
Data augmentation [30] increases dataset size and diversity by applying transformations such as flipping, rotating, scaling, cropping, translating, and bright-

ness or noise adjustments. It improves model generalization, reduces overfitting, and enhances robustness to input variations. In practice, libraries like `Keras` offer tools such as `ImageDataGenerator` to automate augmentation during training.

5.4 Pooling

Pooling is an operation commonly used in CNNs to reduce the spatial dimensions of feature maps while retaining important features. Pooling layers operate independently on each feature map and aggregate information within local neighborhoods. The most commonly used pooling methods in convolutional neural networks are max pooling and average pooling. Max pooling selects the highest value from each sub-region of the feature map. For example, given a 2×2 sub-region:

$$\text{Input:} \quad \begin{bmatrix} 1 & 3 \\ 2 & 4 \end{bmatrix} \quad \Rightarrow \quad \text{Max pooled output:} \quad \max(1, 3, 2, 4) = 4$$

Average pooling computes the mean value of each sub-region, resulting in a smoothed representation. For example, given the following 2×2 sub-region:

$$\text{Input:} \quad \begin{bmatrix} 1 & 3 \\ 2 & 4 \end{bmatrix} \quad \Rightarrow \quad \text{Average pooled output:} \quad \frac{1+3+2+4}{4} = 2.5$$

The primary purpose of pooling is to provide spatial invariance, meaning the network can recognize patterns regardless of their precise location within the input. Max pooling, being more common, emphasizes the most prominent features, which makes it useful for tasks where the presence of a feature is more important than its exact value or location.

5.5 VGG-19

VGG-19 [31] is a widely used deep CNN known for its simple and uniform architecture, making it a standard baseline for image classification tasks. Figure 5.2 illustrates the structure of VGG-19. The model consists of 16 convolutional layers grouped into 5 sequential blocks. Each block is followed by a max pooling layer that reduces the spatial resolution of the feature maps by a factor of 2 while typically doubling the number of filters. After the convolutional blocks, the network includes three fully connected (dense) layers with 4096, 4096, and 1000 units, respectively. The original

Fig. 5.2 Architecture of VGG-19. An input image is processed from bottom to top through a sequence of convolutional blocks with small 3×3 filters and max-pooling layers, followed by fully connected layers and a final softmax classifier

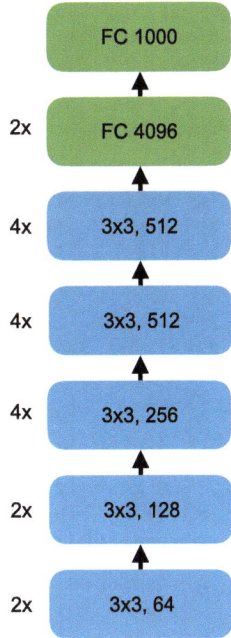

VGG-19 model was trained on the ImageNet dataset and classifies input images into 1000 distinct classes.

ResNet

ResNet (Residual Network) [12] is a deep learning architecture that enables training of very deep networks by using skip connections. Instead of learning the full mapping, each block learns a residual function:

$$y = \mathcal{F}(x) + x$$

where $\mathcal{F}(x)$ is the transformation and x is added via a shortcut. This helps preserve information and improves gradient flow, making deep models like ResNet-18 or ResNet-50 stable and effective.

5.6 CNN in Keras

CNNs can be implemented in Keras for image classification tasks like CIFAR-10, a dataset of 60,000 32×32 color images across 10 classes, including airplanes, automobiles, birds, cats, deer, dogs, frogs, horses, ships, and trucks.

```
from keras.models import Sequential
from keras.layers import Conv2D, MaxPooling2D, Flatten, Dense,
    Dropout
from keras.datasets import cifar10
from keras.utils import to_categorical

# Load and preprocess CIFAR-10
(X_train, y_train), (X_test, y_test) = cifar10.load_data()
X_train, X_test = X_train / 255.0, X_test / 255.0
y_train, y_test = to_categorical(y_train),
    to_categorical(y_test)

# Build the CNN
model = Sequential([
    Conv2D(32, (3, 3), activation='relu', input_shape=(32, 32,
        3)),
    MaxPooling2D(pool_size=(2, 2)),
    Conv2D(64, (3, 3), activation='relu'),
    MaxPooling2D(pool_size=(2, 2)),
    Conv2D(128, (3, 3), activation='relu'),
    MaxPooling2D(pool_size=(2, 2)),
    Flatten(),
    Dense(256, activation='relu'),
    Dropout(0.5),
    Dense(10, activation='softmax')])

# Compile, train, and evaluate the model
model.compile(optimizer='sgd', loss='categorical_crossentropy',
    metrics=['accuracy'])
model.fit(X_train, y_train, validation_data=(X_test, y_test),
    epochs=25, batch_size=64)
test_loss, test_acc = model.evaluate(X_test, y_test)
print("Test accuracy:", test_acc)
```

At the beginning, the images are normalized. The CNN includes convolutional layers with increasing filters (32, 64, 128) using ReLU, pooling layers for down-sampling, a dense layer with 256 units, and dropout to prevent overfitting. It uses the SGD optimizer and categorical cross-entropy loss for training. The architecture can be tuned for improved performance.

Deepness in Deep Learning
Deep neural networks are formed by stacking layers such as dense and convolutional layers, often combined with pooling, normalization, and activation functions. As depth grows, the network learns more abstract features. Early layers detect edges, while deeper layers detect object parts and entire objects. Architectures like VGG-16 (16 layers) and ResNet-152 (152 layers) show how depth can improve performance. Residual connections stabilize training in very deep networks and allow models with more than 1000 layers. Such depth, however, requires large amounts of data.

5.7 Exercises

The following exercises help you practice key concepts from this chapter, including convolution, pooling, and CNN training.

- Questions:

  - What is a convolution in CNNs, and why is it useful for images?
  - What is the main idea behind GoogLeNet [33]?
  - What do pooling layers do, and how do they help reduce computation and extract features?

- Hands on:

  - Use the following 3×3 input and 2×2 filter to do a 2D convolution with stride 1 and no padding.
    Input:

    $$\begin{bmatrix} 2 & 1 & 0 \\ & 3 & 2 \\ & 1 & 1 \end{bmatrix} \quad \text{Filter: } \begin{bmatrix} 0 & 1 \\ & -1 \end{bmatrix}$$

  - Slide the filter over the input and calculate the result for each position. Write down the final 2×2 output matrix.

- Programming:

  - Use the Google Colab notebook to train a simple CNN on the CIFAR-10 dataset. Try changing the learning rate, number of filters, or training time to improve accuracy.
  - Add data augmentation (like flipping or rotating images) and dropout layers to avoid overfitting. Compare the accuracy before and after.
  - Load a pre-trained CNN like VGG or ResNet from `Keras`, upload your own image, and classify it. Optionally, fine-tune the model for better results.

MIT Moral Machine
Visit the MIT Moral Machine and step through a few ethical dilemmas. Sketch the choices you made and note any patterns or surprises. Then, jot down how those patterns reveal your own biases and consider one way designers could reduce bias in such systems.

Chapter 6
Attention

6.1 Sequence Learning

Models based on attention, such as transformers [36] , changed how we process sequences. Unlike older recurrent neural networks that handle sequences step by step, transformers use attention to examine the entire sequence at once and do so much faster. Recurrent models often struggle with long-term dependencies because they process tokens in sequence. Transformers avoid this limit. They model relations between any two tokens, no matter how far apart they are. This design improves efficiency and accuracy in many NLP and sequence tasks. Today, transformers are a core tool in deep learning.

> **Attention Intuition**
> In the sentence AI is fun, the word fun may assign strong attention to AI. The final vector for fun will then include information about AI. This helps the model understand their connection. Attention lets each word view the full sentence and extract useful context.

6.2 Attention

A transformer processes the sentence:

```
AI is fun.
```

The model first splits the sentence into tokens:

```
[AI, is, fun].
```

© The Author(s), under exclusive license to Springer Nature Switzerland AG 2026
O. Kramer, *Artificial Intelligence Essentials*,
https://doi.org/10.1007/978-3-032-06637-4_6

Each token is then converted into a vector. This step is called embedding. It lets the model treat words as numerical objects. The vectors may be learned from scratch or initialized from pre-trained systems such as `word2vec` or `GloVe`. Transformer models usually learn embeddings end-to-end. The embeddings are contextual and depend on position and surrounding tokens. For simplicity, assume we use 2D vectors. The tokens map to:

`AI = [1, 0],   is = [0, 1],   fun = [1, 1]`

These vectors form the matrix:

$$\mathbf{X} = \begin{bmatrix} 1 & 0 \\ 0 & 1 \\ 1 & 1 \end{bmatrix} \begin{matrix} \text{AI} \\ \text{is} \\ \text{fun} \end{matrix}$$

In real transformer models, each word is mapped to a high-dimensional vector such as 512 or 768 dimensions. The embeddings are trained so that words with similar usage receive similar vectors. Transformers do not encode order by default because attention is permutation-invariant. To give the model positional information, a positional encoding vector is added:

$$\mathbf{X}_{\text{pos}} = \mathbf{X} + \text{PositionalEncoding}$$

This lets the model distinguish token positions such as first, second, and third. Each input vector is transformed into a query ($\mathbf{Q}$), key ($\mathbf{K}$), and value ($\mathbf{V}$) using learned weight matrices:

$$\mathbf{Q} = \mathbf{X}_{\text{pos}}\mathbf{W}^Q, \quad \mathbf{K} = \mathbf{X}_{\text{pos}}\mathbf{W}^K, \quad \mathbf{V} = \mathbf{X}_{\text{pos}}\mathbf{W}^V$$

Positional Encoding
Transformers have no built-in sense of order. A positional encoding is added to each token vector to provide this information. A common version uses fixed sinusoidal functions with different wavelengths, creating smooth periodic patterns across embedding dimensions. These patterns let the model capture absolute and relative positions. Positional encodings may also be learned. In both cases, they give the model a reference for sequence structure.

The three projections give different views of the same input. The query vector represents what the token is looking for. The key vector represents what the token offers. The value vector contains the information the token contributes. The model compares each query to all keys using a dot product:

$$\mathbf{Q}\mathbf{K}^\top$$

This produces similarity scores that show how strongly each token relates to others. These scores are scaled and passed through a softmax:

$$\text{softmax}\left(\frac{\mathbf{Q}\mathbf{K}^\top}{\sqrt{d_k}}\right)$$

The softmax converts scores into probabilities. These probabilities weight the value vectors:

$$\text{Attention}(\mathbf{Q}, \mathbf{K}, \mathbf{V}) = \text{softmax}\left(\frac{\mathbf{Q}\mathbf{K}^\top}{\sqrt{d_k}}\right)\mathbf{V}$$

This produces a new vector for each token. Each vector mixes the token's own content with information from tokens it attends to.

Multi-Head Attention
Multi-head attention computes several attention operations in parallel. Each head uses its own learned projections. The outputs are concatenated and then linearly transformed. Each head can focus on a different aspect of the input, such as syntax, semantics, or positional relations. By attending to several types of information at once, the model captures a richer set of patterns in the sequence.

6.3   Hands On Attention

To see how self-attention works, we walk through a simple example without positional encoding. Consider the sentence: Robots learn fast. After tokenization, we obtain three tokens. Each token is mapped to a 2D embedding:

$$\mathbf{X} = \begin{bmatrix} 1 & 1 \\ 1 & 0 \\ 0 & 1 \end{bmatrix}.$$

We compute queries $\mathbf{Q}$, keys $\mathbf{K}$, and values $\mathbf{V}$ by multiplying $\mathbf{X}$ with learned weight matrices:

$$\mathbf{W}^Q = \begin{bmatrix} 2 & 0 \\ 0 & 1 \end{bmatrix}, \quad \mathbf{W}^K = \begin{bmatrix} 1 & 1 \\ 1 & -1 \end{bmatrix}, \quad \mathbf{W}^V = \begin{bmatrix} 1 & 0 \\ 0 & 3 \end{bmatrix}.$$

Applying these gives

$$\mathbf{Q} = \mathbf{X}\mathbf{W}^Q = \begin{bmatrix} 2 & 1 \\ 2 & 0 \\ 0 & 1 \end{bmatrix}, \quad \mathbf{K} = \mathbf{X}\mathbf{W}^K = \begin{bmatrix} 2 & 0 \\ 1 & 1 \\ 1 & -1 \end{bmatrix},$$

$$\mathbf{V} = \mathbf{X}\mathbf{W}^V = \begin{bmatrix} 1 & 3 \\ 1 & 0 \\ 0 & 3 \end{bmatrix}.$$

Next, we compute the scaled dot-product attention scores:

$$\mathbf{Q}\mathbf{K}^\top = \begin{bmatrix} 4 & 3 & 1 \\ 4 & 2 & 2 \\ 0 & 1 & -1 \end{bmatrix}, \quad \frac{1}{\sqrt{2}}\mathbf{Q}\mathbf{K}^\top \approx \begin{bmatrix} 2.83 & 2.12 & 0.71 \\ 2.83 & 1.41 & 1.41 \\ 0.00 & 0.71 & -0.71 \end{bmatrix}.$$

We apply softmax row-wise to obtain the attention weights:

$$\mathrm{softmax}\left(\frac{\mathbf{Q}\mathbf{K}^\top}{\sqrt{2}}\right) \approx \begin{bmatrix} 0.620 & 0.306 & 0.074 \\ 0.673 & 0.164 & 0.164 \\ 0.284 & 0.576 & 0.140 \end{bmatrix}.$$

Finally, we compute the attention output by multiplying the weight matrix with the values:

$$\mathrm{Attention}(\mathbf{Q}, \mathbf{K}, \mathbf{V}) = \mathrm{softmax}\left(\frac{\mathbf{Q}\mathbf{K}^\top}{\sqrt{2}}\right)\mathbf{V} \approx \begin{bmatrix} 0.926 & 2.083 \\ 0.836 & 2.509 \\ 0.860 & 1.272 \end{bmatrix}.$$

Each row is the new contextual vector for a token. It combines its own content with information from other tokens based on the learned attention weights.

Layer Normalization
Layer normalization [2] stabilizes training by normalizing values in each input vector across its features. It computes mean and standard deviation, rescales values to zero mean and unit variance, and then applies learned scale and shift parameters. This keeps activations within a stable range, improves gradient flow, and enhances training stability and generalization.

6.4   GPT

GPT (Generative Pre-trained Transformer) [24] is a language model built on the decoder-only transformer architecture. It generates text autoregressively by predicting the next token based on all previous tokens:

Fig. 6.1 Architecture of a
Transformer-based language
model (GPT-style). Input
tokens are mapped to text
and positional embeddings,
processed through stacked
self-attention blocks with
multi-head attention, layer
normalization, and
feed-forward (dense) layers,
and finally projected to
predict the next token

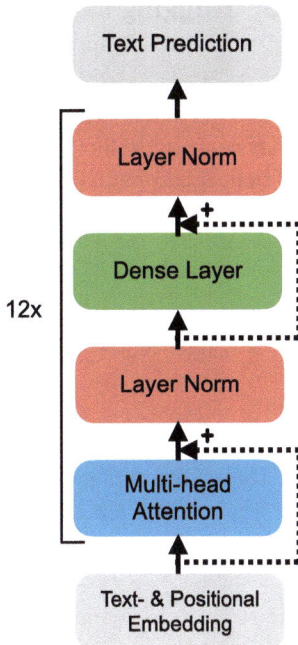

$$P(x_1, x_2, \ldots, x_n) = \prod_{t=1}^{n} P(x_t \mid x_1, \ldots, x_{t-1})$$

GPT uses masked self-attention so each token can only attend to earlier tokens during training and inference. Each token is embedded into a vector, and positional encoding is added to provide order. The vectors pass through a stack of decoder blocks, see Fig. 6.1. Each block contains masked attention, feedforward layers, and layer normalization.

GPT is pre-trained on large text corpora and can be fine-tuned for specific tasks. Given a prompt such as:

```
AI is fun because it ____
```

GPT may complete it with phrases like `can generate stories`, `learns from data`, or `helps solve problems`.

Transformers
Transformers [36] changed sequence modeling and serve as the base of many state-of-the-art models. They consist of an encoder and a decoder built from blocks containing multi-head attention, layer normalization, dense layers, and residual connections. GPT [24] uses only the decoder for autoregressive generation. BERT [8] uses only the encoder for bidirectional context understanding.

6.5   Hugging Face

Hugging Face's `transformers` library provides simple access to many pre-trained models for tasks such as classification, translation, summarization, and sentiment analysis. The core tool is the `pipeline`, which handles model loading, tokenization, inference, and output formatting in a few lines of code. Below is an example for sentiment analysis:

```python
from transformers import pipeline

model_name = "distilbert-base-uncased-finetuned-sst-2-english"
sentiment_pipeline = pipeline("sentiment-analysis",
    model=model_name)

text = "I love using Hugging Face's tools!"
result = sentiment_pipeline(text)
```

This code uses a pre-trained model through the pipeline. A possible output is:

```
Sentiment: POSITIVE, Confidence: 0.99
```

Streamlit

Streamlit is a Python framework for building interactive web apps with minimal code. It is used for data science demos, machine learning dashboards, and educational tools. A Streamlit app behaves like a Python script that runs top-to-bottom on each interaction. A simple deployment concept make it ideal for fast prototyping and sharing of AI experiments.

6.6   Exercises

These exercises help you practice the core ideas of transformers, attention, positional encoding, and their applications in modern NLP.

- Questions:

  - What are the building blocks of the transformer architecture?
  - What is the difference between multi-head and single-head attention?
  - Why do transformers need positional encoding?
  - What does autoregressive mean in the context of GPT?

- Hands on:

  - Take the following matrices for a simple attention example with three tokens:

$$\mathbf{Q} = \begin{bmatrix} 1 & 0 \\ 1 & 1 \\ 0 & 1 \end{bmatrix}, \quad \mathbf{K} = \begin{bmatrix} 1 & 0 \\ 1 & 1 \\ 0 & 1 \end{bmatrix}, \quad \mathbf{V} = \begin{bmatrix} 5 & 0 \\ 2 & 2 \\ 0 & 5 \end{bmatrix}$$

  - Compute the attention output.
  - Look at the result and describe how much each token attends to the others.

- Programming:

  - Build a simple Streamlit app that lets a user type a sentence and shows the predicted sentiment (e.g. positive, negative, neutral) using a model from Hugging Face. Also display the confidence score.
  - Explore other Hugging Face models like Named Entity Recognition (NER). Try out the model dslim/bert-base-NER and report what it finds in some example sentences.

Discover AI Conferences
Explore AI conferences such as ESANN, NeurIPS, ICML, or ICLR.

Part III
LLMs

Chapter 7
Large Language Models

7.1 LLMs

Large Language Models (LLMs) are AI systems that read, understand, and generate human language. They are built with transformers and trained on large text corpora. This training lets them learn grammar, facts, and simple reasoning patterns. LLMs turn input text into numerical units called tokens. The model processes these tokens internally and then generates one token at a time to form a response. The output tokens are converted back into readable text. A tokenizer performs this mapping. It splits text into words, subwords, or characters and assigns numerical values the model can process. LLMs are often trained as foundational models. They are not designed for one fixed task but can adapt to many. After broad training, they can be fine-tuned or guided with examples for tasks such as question answering, summarization, or story writing. State-of-the-art LLMs are called frontier models. They define current AI capability but require high computational and energy resources to train.

7.2 Parameters and Precision

The number of parameters influences the capability of LLMs. Models with more parameters can typically represent more complex patterns and achieve stronger generalization, while smaller models offer advantages in speed, efficiency, and deployment cost. However, increasing parameter counts also raises computational requirements for both training and inference. Modern LLM families therefore span a wide spectrum of sizes, from lightweight variants optimized for fast on-device use to large models designed for high-capacity reasoning and generation. Different scales offer different trade-offs between computational efficiency, memory footprint, and output quality, allowing practitioners to choose model sizes that best match their application and resource constraints. Numerical precision defines how values are stored inside a

O. Kramer, *Artificial Intelligence Essentials*,
https://doi.org/10.1007/978-3-032-06637-4_7

model. FP32 is the usual choice for training because it yields stable gradients. FP16 reduces memory use and improves speed with only minor accuracy loss. INT8 and INT4 shrink memory needs further and are mainly used for inference on restricted hardware. Such low-precision formats make it possible to run large models on devices with limited resources.

7.3   Ollama

Ollama is available for macOS, Linux, and Windows. It provides a lightweight framework for running language models locally, such as Llama [34], Gemma, and Mistral. It offers a simple API to load, run, and manage models and includes many ready-to-use models. The commands below show how to list, download, and run models.

```
# List available models
ollama list
```

```
# Pull a specific model (e.g., Llama 3.3)
ollama pull llama3
```

```
# Run the model
ollama run llama3
```

For scripted use, text can be passed directly through standard input. Ollama models can be customized with a Modelfile. The file defines the base model, parameters, and system behavior. This allows control over instructions, quantization, and tokenization.

```
FROM mistral
PARAMETER temperature 0.7

SYSTEM "You are a helpful assistant specialized in answering
    technical questions."
```

This configuration loads the mistral model, sets the temperature to 0.7, and defines a simple system prompt. Ollama enables flexible local deployment while keeping full control on the device.

Vision Models
Ollama supports multimodal language models such as Llama 3.2 Vision. These models process text and images together. They handle tasks like image recognition, OCR, and image-based question answering. When an image is sent with a prompt, the model can describe the scene, extract text, or answer visual questions. This combines language processing with visual perception.

7.4 Prompting

Prompting is the main interface between users and LLMs. The phrasing of a question or instruction often shapes the quality of the output. In zero-shot prompting, the model receives only the task description. This works well for simple tasks but can fail on complex or multi-step problems. An example is:

```
Q: A bird flew 6 kilometers in the morning and 4 kilometers in the evening.
   How far did the bird fly in total?
```

In one-shot prompting, a single example is provided to demonstrate the desired input-output behavior. This helps the model imitate a reasoning pattern and improves the consistency and reliability of its responses:

```
Q: A car travels 10 kilometers in the morning and 5 kilometers in the
   evening. How far does the car travel in total?
A: 10 + 5 = 15 kilometers

Q: A bird flew 6 kilometers in the morning and 4 kilometers in the evening.
   How far did the bird fly in total?
```

In few-shot prompting, multiple examples are provided before the main question. With more demonstrations, the model can better generalize the task pattern, especially for more complex or unfamiliar inputs:

```
Q: A car travels 10 kilometers in the morning and 5 kilometers in the
   evening. How far does the car travel in total?
A: 10 + 5 = 15 kilometers

Q: A train covers 50 kilometers before lunch and 30 kilometers after lunch.
   How far does the train travel in total?
A: 50 + 30 = 80 kilometers

Q: A bird flew 6 kilometers in the morning and 4 kilometers in the evening.
   How far did the bird fly in total?
```

Chain of Thought (CoT) prompting [37] enhances few-shot prompting by including intermediate reasoning steps. Instead of directly producing an answer, the model is guided to solve the problem step by step. This technique improves accuracy, interpretability, and robustness, especially in tasks requiring logical deduction, multi-step mathematics, or scientific analysis.

```
Q: A bakery sells 3 types of muffins. Each muffin costs $2, and a customer
   buys 5 of each type. How much does the customer pay?

Step-by-step solution:
1. The customer buys 5 muffins of each type.
2. There are 3 types, so the total muffins purchased is 5 x 3 = 15.
3. Each muffin costs $2, so the total cost is 15 x 2 = 30.
4. The final answer is $30.

Now solve a similar problem using step-by-step reasoning:

Q: A bookstore sells 4 types of books. Each book costs $5, and a customer
   buys 6 of each type. How much does the customer pay?
```

By guiding the model through a staged reasoning process, Chain of Thought prompting produces clearer and more accurate results in complex cases.

> **GAIA Benchmark**
> The GAIA Benchmark (General AI Assessment) [22] evaluates reasoning, generalization, and adaptability across tasks involving language, vision, math, code, and knowledge. It measures broad capability rather than narrow skill, making it a useful tool for tracking progress toward general intelligence.

7.5   A Chatbot in `Ollama`

The following example demonstrates a minimal chatbot using Ollama.

```python
import requests, json
model = "llama3.2"

def chat(messages):
    r = requests.post("http://localhost:11434/api/chat",
                      json={"model": model, "messages":
                            messages, "stream": True},
                      stream=True)
    output = ""
    for line in r.iter_lines():
        msg = json.loads(line)
        if "error" in msg:
            raise Exception(msg["error"])
        if not msg.get("done"):
            content = msg.get("message", {}).get("content", "")
            output += content
            print(content, end="", flush=True)
    return {"role": "assistant", "content": output}

messages = [{"role": "system", "content": "Answer in max three
    sentences, concise and ironic."}]

while True:
    user = input("You: ")
    if not user: break
    messages.append({"role": "user", "content": user})
    messages.append(chat(messages))
    print("\n")
```

This minimal script implements a conversational interface with a locally running LLaMA model via the Ollama API. It sends user input to the model and streams the response back in real time using `stream=True`. The core of the communication is a POST request to `http://localhost:11434/api/chat`, which includes a `messages` parameter, a list of dictionaries like `{"role": "user", "content": "Hello!"}`. This list represents the full dialogue history and allows the model to maintain conversational context across turns. The payload is encoded in JSON (JavaScript Object Notation), a lightweight data format commonly used to exchange structured data between applications.

Reinforcement Learning with Human Feedback
Reinforcement Learning with Human Feedback (RLHF) [5] is a training method used to align LLMs with human preferences. Instead of optimizing only for accuracy or likelihood of the next token, RLHF teaches the model to generate responses that humans find helpful, safe, or aligned with specific goals. The process typically involves three steps: (1) collecting human feedback on model outputs, (2) training a reward model to predict which outputs are preferred, and (3) fine-tuning the LLM using reinforcement learning to generate responses that score higher according to the reward model.

7.6 Exercises

Review and apply the key concepts from this chapter on LLMs, prompting, and deployment.

- Questions:

  - What is the difference between foundational and frontier models?
  - Why are INT8 and INT4 useful for inference?
  - What can a vision LLM do?
  - How does Ollama help run models locally?
  - What is the idea of Chain of Thought prompting?

- Programming:

  - Build a basic chatbot using an LLM and Streamlit.
  - Add a sentiment model and adapt responses.
  - Compare answers to math questions with and without Chain of Thought prompts.

AI-Generated Music
Generate music with AI. Which are your favorite tools?

Chapter 8
Agentic AI

8.1 Agents

Agentic AI refers to systems that show autonomous, goal-directed behavior in changing environments. Traditional models process inputs passively. Agentic systems act, plan, and adapt with minimal human intervention. Key traits include self-set goals, real-time planning, fast adaptation, and strategy updates based on feedback. LLMs are not agentic by default. They become agentic when placed inside agent frameworks that provide memory, explicit reasoning steps, tool usage, and links to external systems. These architectures let LLM-based agents solve complex tasks, refine their actions, and coordinate subtasks. Figure 8.1 shows an overview.

> **Multi-Agent Systems**
> A multi-agent system (MAS) contains several autonomous agents that act in a shared environment. Each agent follows its own goal and interacts through simple rules. System behavior emerges from these interactions. LLM-based agents communicate, plan, and coordinate through natural language. Roles such as planner, executor, or evaluator structure the workflow. MAS enable complex tasks through distributed cooperation.

8.2 Agent Memory

Memory is a central element of agentic behavior. It lets an agent store past interactions, track outputs, review tool results, and adjust its strategy. Without memory, an LLM-based agent is stateless and cannot reflect on earlier actions. Many current architectures implement memory through prompt engineering. Past information is

Fig. 8.1 An LLM agent combines a language model with structured prompts for reasoning, planning, and acting, supported by memory and external tools for adaptive task execution

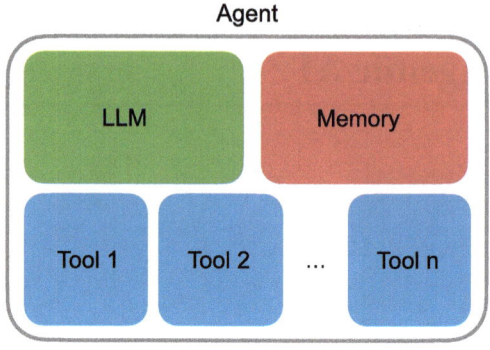

inserted dynamically into the model's context window. This mirrors the classic blackboard architecture from symbolic AI. Different modules read and write to a shared memory space, enabling collaborative problem solving, multi-step reasoning, and iterative refinement. Memory also enables consistency checks. Outputs from different reasoning paths can be compared and reconciled with voting, meta-reasoning, or fallback rules.

Retrieval-Augmented Generation

Retrieval-Augmented Generation (RAG) [20] grounds model outputs in external documents such as manuals or research papers. LLMs store general knowledge, but RAG fetches specific information for more accurate answers. Each document is converted into an embedding:

$$\text{doc\_vec}_i = \text{encode}(\text{doc}_i)$$

A user question is embedded in the same way:

$$\text{query\_vec} = \text{encode}(\text{query})$$

The system retrieves the top-k documents whose embeddings are most similar to the query vector (typically via cosine similarity). The retrieved texts are then inserted into the prompt:

```
[CONTEXT: doc₁,...,docₖ]  [QUERY: query]
```

The model uses this context to generate a better answer. This reduces hallucination and improves factual accuracy.

8.3 Agent Tools

Tools expand what an agent can do. They give a language model access to the external world. A model can generate text, but a tool lets it act: browse the web, execute code, read files, analyze images, or cooperate with other agents. A tool is an external function the agent can call when needed.

Tools turn static models into interactive systems. An agent might search online, extract content from a PDF, run code to analyze the results, and summarize the findings. Frameworks like LangChain, CrewAI, and AutoGen define how agents call tools, read outputs, and link these steps into reasoning chains. Tools can be local, such as a code interpreter, or remote, such as a web API. Typical tasks include file reading, math or plotting, transcription, web search, and multi-agent coordination. With context-based tool selection, agents become modular and adaptive across many tasks.

OpenAI Whisper

Whisper [25] is a speech-to-text model by OpenAI. It transcribes spoken language into written text. Trained on broad multilingual data, it handles noisy audio, many languages, and simple translation. As a tool for an agent, it listens to audio, processes it with transformer models, and outputs text or translations.

8.4 Agents in CrewAI

CrewAI is a lightweight Python framework for building agentic AI systems from modular agents and tasks. You define agents with roles and goals, assign tasks, and let the runtime coordinate their interaction. The next example shows a minimal multi-agent setup where one agent asks a question and another agent answers it with GPT-4. To install CrewAI and the required integration, run:

```
pip install crewai langchain langchain-openai
```

The code below creates two agents and two tasks, one for generating a question and one for providing an answer:

```python
import os
from crewai import Agent, Task, Crew
from langchain_openai import ChatOpenAI

# Set up the OpenAI LLM
os.environ["OPENAI_API_KEY"] = "your-api-key"
llm = ChatOpenAI(model="gpt-4", temperature=0.7)
```

```
# Agent 1: Student asks a question
student = Agent(
    role="Curious Student",
    goal="Ask one specific, insightful question about AI",
    backstory="A learner eager to probe deep AI topics",
    llm=llm
)
ask_q = Task(
    description="Ask one insightful question about artificial
        intelligence.",
    expected_output="A single, thought-provoking AI question.",
    agent=student
)

# Agent 2: Expert answers the question
expert = Agent(
    role="AI Expert",
    goal="Provide a clear, deep answer",
    backstory="A professor guiding students through AI
        concepts",
    llm=llm
)
answer_q = Task(
    description="Answer the student's AI question clearly.",
    expected_output="A concise, comprehensive answer to the
        question.",
    agent=expert
)

# Execute the workflow
crew = Crew(
    agents=[student, expert],
    tasks=[ask_q, answer_q],
    llm=llm,
    verbose=False   # Set to True for step-by-step logs
)

result = crew.kickoff()
print("Final Answer:", result)
```

This example illustrates the modular and conversational nature of CrewAI. Agents can reason sequentially, passing implicit context between tasks. By chaining roles, you can orchestrate multi-step workflows using language alone. In more advanced scenarios, you can integrate external tools, long-term memory, or custom evaluation logic.

Turing Test

The Turing Test [35], proposed by Alan Turing in 1950, is a benchmark for judging machine intelligence. A human interacts with a machine and another human through text only. If the judge cannot reliably tell which is which, the machine is considered to have passed the test. In that case, its behavior appears indistinguishable from human intelligence.

Table 8.1 LLM-Based Agent Frameworks

Name	Description
LangChain	Chains tools and reasoning into LLM-based agents
LangGraph	Graph-based control flow for agent memory and state
AutoGPT	Autonomous agent that breaks down and executes goals
CrewAI	Role-based multi-agent system with task planning

8.5 LLM-Based Agent Frameworks

The development of agentic AI has given rise to a wide variety of frameworks and systems that support autonomous behavior in large language models. These systems differ in scope, structure, and specialization, but they share the common goal of enabling LLMs to act purposefully in dynamic environments. Table 8.1 provides an overview of popular LLM-based agent frameworks. These include general-purpose libraries such as LangChain Agents and LangGraph , iterative goal-driven agents like AutoGPT, and structured multi-agent platforms like CrewAI. Each framework enables LLMs to plan, coordinate, and act with varying degrees of memory, tool usage, and role-based interaction.

> **Artificial General Intelligence**
> Artificial General Intelligence (AGI) [9] describes AI systems with human-level flexibility: the ability to learn, understand, reason, and solve problems across a wide range of tasks. Unlike narrow AI, AGI can generalize, adapt to new situations, and acquire new skills autonomously. While still theoretical, AGI is a major research goal with active debate around its potential, risks, and timeline.

8.6 Exercises

This set of exercises explores the ideas behind LLM agents, memory, and tool use.

- Questions:
  - What is an LLM agent, and how is it different from a standard language model?
  - How does memory help an agent perform better?
  - Why are tools important in agentic AI?

- Programming:

  - Build an agent pipeline with CrewAI, LangGraph, or another agent framework.
  - Use agents with different roles. For example, one agent fetches a Wikipedia article, and another creates a beginner-friendly summary.
  - Test tools such as Whisper, Ollama, and Hugging Face in an agent setup.
  - Ensure the system passes information between agents and produces a clear output.

Humanoid Robots
Find out about what the most advanced humanoid robots are, e.g. G1 and Optimus.

Part IV
Optimization

Chapter 9
Genetic Algorithms

9.1 Optimization

Optimization is the task of finding the best solution among many candidates by maximizing or minimizing an objective function. This function measures the quality or performance of each solution. Optimization is central in machine learning, engineering, and science whenever a system must make an optimal decision or choose an optimal configuration. A simple numerical example is the SPHERE function in Fig. 9.1. The function assigns lower values to better solutions, with the minimum at the origin.

To build intuition, consider the symbolic task of evolving the target string GENETIC ALGORITHMS. Here, the goal is to match a fixed target. The objective function, or fitness function, counts how many characters are correct. Objective functions in other domains may be numerical and continuous. For example, linear regression finds a best-fit line by minimizing the mean squared error. Many real-world problems allow evaluation of a solution but do not offer an analytic or differentiable form. These are black-box optimization problems.

> **Biological Evolution**
> Genetic algorithms draw inspiration from Darwin's theory of natural selection. Variation and survival drive adaptation over generations. Mutation and recombination create diversity, and individuals with stronger traits reproduce more often. Over time, populations become better suited to their environment. This idea transfers to computational search.

© The Author(s), under exclusive license to Springer Nature Switzerland AG 2026
O. Kramer, *Artificial Intelligence Essentials*,
https://doi.org/10.1007/978-3-032-06637-4_9

Fig. 9.1 Sphere function $f(x_1, x_2) = x_1^2 + x_2^2$ shown as a 3D surface, representing a simple numerical optimization problem with a single global minimum

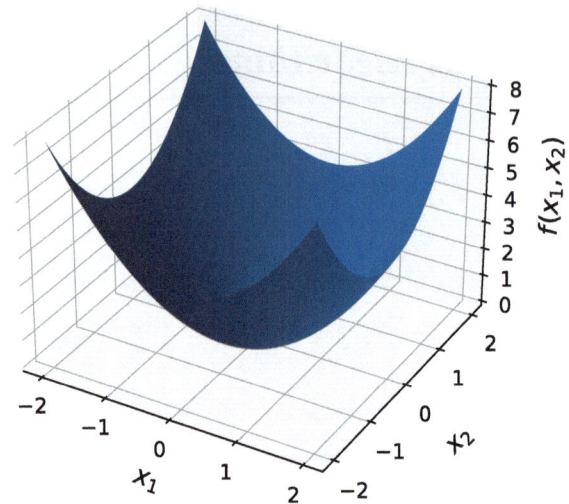

9.2 Evolution

Genetic Algorithms (GAs) [14, 29] convert Darwinian evolution into computational optimization. Each individual represents a candidate solution, called a chromosome. It may use a binary string, a list of numbers, or a set of symbols as its encoding. The internal representation is the genotype. The expressed solution in the problem space is the phenotype. A fitness function evaluates each individual's quality. The algorithm evolves a population through repeated cycles of crossover, mutation, fitness evaluation, and selection, shown in Fig. 9.2 [16].

A common variant is the $(\mu + \lambda)$-GA [3]. A population of μ parents produces λ offspring through crossover and mutation. After evaluating all individuals, the algorithm selects the best μ candidates from the combined set of parents and offspring. Strong solutions persist, while new ones maintain diversity. The (μ, λ)-GA differs by selecting only from the offspring. This increases exploration but may discard strong

Fig. 9.2 Evolutionary cycle: crossover, mutation, fitness evaluation, and selection

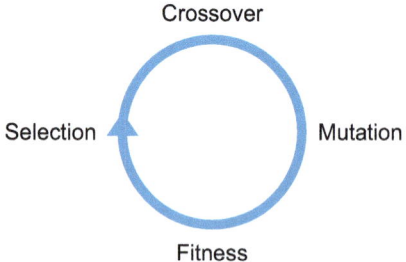

parents. The algorithm stops when a chosen condition is met, such as a fixed number of generations or a target fitness value.

9.3 Crossover

Crossover is a genetic operator that combines parts of two parent solutions to create new offspring. It mimics sexual reproduction by mixing information from both parents. This promotes diversity and allows useful traits to appear together in one solution. In string-based GAs, such as evolving the phrase GENETIC ALGORITHMS, each individual is a sequence of characters. A simple method is single-point crossover. A random index is chosen, and the offspring is formed by taking the prefix from one parent and the suffix from the other. For example:

Parent 1: GENE|TIC ALGORITHMS   Parent 2: EVOL|VING SOLUTIONS

Child: GENEVING SOLUTIONS

This type of recombination helps useful substrings, such as correct partial matches, spread through the population over generations. A simple Python version of single-point crossover is shown below:

```
def crossover(parent1, parent2):
    idx = random.randint(0, len(parent1) - 1)
    return parent1[:idx] + parent2[idx:]
```

Although simple, this operator plays a key role in the search process. Over many generations, crossover combines useful character sequences and increases the chance of reaching the target string. The same idea applies to more complex GAs in numerical optimization or machine learning, where crossover acts on binary strings or real-valued vectors.

9.4 Mutation

Mutation introduces random variation into the population and maintains genetic diversity. In string-based GAs such as evolving the phrase GENETIC ALGORITHMS, mutation randomly changes characters in a string. A common method is character mutation. Each character has a small chance of being replaced with another character from the allowed set, such as uppercase letters and spaces. For example:

GENETIC ALGORITHMS   becomes   GENETIC ALPORITHMS

Here, one character in the second word was changed to P. Most characters remain unchanged, but occasional edits help the population explore new possibilities and escape local optima. Mutation becomes essential when the population grows too similar. Without it, the algorithm may converge early to a weak solution. The mutation rate is usually low (often between 1% and 10%) to preserve good solutions while still enabling exploration. A simple Python version for string mutation is:

```
def mutate(individual, rate=0.1):
    return ''.join(
        c if random.random() > rate else
            random.choice("ABCDEFGHIJKLMNOPQRSTUVWXYZ ")
        for c in individual
    )
```

This function replaces characters at random based on the mutation rate. Over many generations, mutation, selection, and crossover work together to move the population toward the optimal string.

Bitflip Mutation

In GAs, individuals are often encoded as fixed-length bitstrings, where each bit represents a gene with a binary value (0 or 1). Bitflip mutation flips single bits with a small probability (typically $p = 1/N$ for bitstrings of length N). For example:

$$101001 \rightarrow 101011$$

Here, the fifth bit was flipped from 0 to 1.

9.5   Hands on GA

We simulate a $(2 + 3)$-GA to evolve the target string GENETIC. Each individual is a 7-character string using uppercase letters. Fitness is the number of characters matching the target in the correct position. Let the initial parent population in Generation 1 be

Parent 1 = GAROTAC   (fitness = 3),   Parent 2 = BENEDIR   (fitness = 4).

Three children are generated using one-point crossover and character-wise mutation. Child 1 is produced by taking the first 4 characters from Parent 1 and the last 3 from Parent 2:

$$\text{GAROTAC}_{[0:4]} \parallel \text{BENEDIR}_{[4:7]} = \text{GARODIR}.$$

Applying mutations R→T and O→E yields

$$\text{Child 1} = \text{GAREDIT}   (\text{fitness} = 3).$$

Child 2 is produced by taking the first 5 characters from Parent 2 and the last 2 from Parent 1:

$$\text{BENEDIR}_{[0:5]} \parallel \text{GAROTAC}_{[5:7]} = \text{BENEDAC}.$$

Applying mutations D→T and A→I yields

$$\text{Child 2} = \text{BENETIC} \quad (\text{fitness} = 6).$$

Child 3 is produced by taking the first 3 characters from Parent 1 and the last 4 from Parent 2:

$$\text{GAROTAC}_{[0:3]} \parallel \text{BENEDIR}_{[3:7]} = \text{GAREDIR}.$$

A normal random mutation might change A→N, I→T, and R→C, giving

$$\text{Child 3} = \text{GNREDTC} \quad (\text{fitness} = 3).$$

After Generation 1, the population is

$$\text{Parent 1} = \text{GAROTAC (3)}$$
$$\text{Parent 2} = \text{BENEDIR (4)}$$
$$\text{Child 1} = \text{GAREDIT (3)}$$
$$\text{Child 2} = \text{BENETIC (6)}$$
$$\text{Child 3} = \text{GNREDTC (3)}$$

Selection chooses the two highest-fitness individuals for the next generation:

$$\text{Parent 1}^{(1)} = \text{BENETIC}, \qquad \text{Parent 2}^{(1)} = \text{BENEDIR}.$$

The GA then continues, gradually improving toward GENETIC.

Neuroevolution
Neuroevolution applies evolutionary algorithms to neural networks. It can optimize architectures, hyperparameters, or weights. A network design can be encoded as a bitstring. Each segment defines layer type, size, activation function, and optional normalization.

Example bitstring:

$$010 \quad 11 \quad 10 \quad 00 \quad 100$$

010	11	10	00	100
Layer type: Conv	Kernels: 8	Activation: ReLU	Norm: None	Next layer: Dense

This genome encodes a convolutional layer with eight kernels, ReLU activation, no normalization, and a dense layer as the next step. Mutation and
crossover act on these codes and evolve stronger architectures over generations.

9.6 GA in `Python`

The next implementation shows a $(10 + 20)$-GA that evolves the target string
`"GENETIC ALGORITHMS"` from random character sequences. It uses random
parent selection, one-point crossover, and character-wise mutation. Each generation keeps the best 10 individuals as parents. They produce 20 offspring. From the
30 candidates, the top 10 form the new parent set.

```python
import random

# Parameters
TARGET = "GENETIC ALGORITHMS"
ALPHABET = "ABCDEFGHIJKLMNOPQRSTUVWXYZ "
MU = 10            # number of parents
LAMBDA = 20        # number of offspring
MUTATION_RATE = 0.1
GENERATIONS = 1000

# Fitness = number of matching characters
def fitness(ind):
    return sum(c1 == c2 for c1, c2 in zip(ind, TARGET))

# Generate a random individual
def random_string():
    return ''.join(random.choice(ALPHABET) for _ in
        range(len(TARGET)))

# 1-point crossover
def crossover(p1, p2):
    idx = random.randint(1, len(TARGET) - 1)
    return p1[:idx] + p2[idx:]

# Mutate a string with per-character probability
def mutate(ind):
    return ''.join(
        c if random.random() > MUTATION_RATE else
            random.choice(ALPHABET)
        for c in ind
    )

# (10+20)-GA
def plus_ga():
    # Initialize population
    population = [random_string() for _ in range(MU)]
    for gen in range(GENERATIONS):
```

```
        # Evaluate and sort by fitness
        population.sort(key=fitness, reverse=True)
        best = population[0]
        print(f"Gen {gen:3d} | Best: {best} (Fitness:
            {fitness(best)})")
        if best == TARGET:
            print("Target reached!")
            return best

        # Generate offspring
        offspring = []
        for _ in range(LAMBDA):
            p1, p2 = random.sample(population, 2)
            child = mutate(crossover(p1, p2))
            offspring.append(child)

        # Plus-selection: combine and select best
        population = sorted(population + offspring,
            key=fitness, reverse=True)[:MU]

    return population[0]

final = plus_ga()
print("\nFinal solution:", final)
```

This symbolic GA shows how simple evolutionary mechanisms can discover structured sequences. The fitness function rewards character matches. Crossover and mutation explore the space of candidate strings. Over time, the population moves toward the target phrase GENETIC ALGORITHMS.

9.7 Exercises

The following exercises help reinforce key concepts from this chapter on GAs using string evolution.

- Questions:

  - What is an optimization problem?
  - How do GAs mimic natural evolution?
  - Describe how crossover and mutation modify strings.
  - What does $(\mu + \lambda)$-GA mean?
  - How does the selection process influence which solutions survive?

- Hands on:

  - Simulate two generations of a $(2 + 3)$-GA that tries to evolve the string GOODBYE.

- Start with the two parents: GUUDPAI and BYEGOOD.
- Use single-point crossover and simple character mutation to generate three children.

- Programming:

  - Write a GA using a population of 20 parents and 100 children to solve the OneMax problem.
  - Represent individuals as bitstrings and define fitness as the number of 1s.
  - Use standard crossover and bit-flip mutation. Run for several generations.
  - Output the best individual found during the run.

AlphaEvolve
Explore Google's AlphaEvolve, an AI system that combines LLMs with evolutionary search to design novel algorithms.

Appendix A
AI Benchmarks

A.1 ML Datasets

Table A.1 provides an overview of several widely used benchmark datasets in machine learning and computer vision.

A.2 LLM Benchmarks

LLMs are evaluated using benchmarks that assess their performance in reasoning and commonsense understanding, see Table A.2.

Table A.1 Overview of standard machine learning datasets

Dataset	Samples	Features (d)	Classes	Task type
Iris	150	4	3	Classification
Wine	178	13	3	Classification
Breast cancer	569	30	2	Classification
Digits	1797	64	10	Classification
Cali. housing	20640	8	–	Regression
MNIST	70000	28×28	10	Image classification
CIFAR-10	60000	$32 \times 32 \times 3$	10	Image classification

O. Kramer, *Artificial Intelligence Essentials*,
https://doi.org/10.1007/978-3-032-06637-4

Table A.2 Selected benchmarks for evaluating LLMs

Benchmark	Description
GSM8K [6]	Grade-school math word problems
MATH500 [13]	Advanced early college-level math problems
GAIA [22]	General AI Alignment benchmark for language models

A.3   Optimization Problems

ONEMAX is defined as:

$$f(\mathbf{x}) = \sum_{i=1}^{N} x_i,$$

where $\mathbf{x} \in \{0, 1\}^N$ is a binary string of length N. The goal is to maximize the number of ones in the bitstring. The global optimum is reached at $\mathbf{x}^* = (1, 1, \ldots, 1)$, with $f(\mathbf{x}^*) = N$. This function is often used as a simple benchmark for evolutionary algorithms, testing their ability to perform bit-level optimization without deception or local optima.

SPHERE is defined as:

$$f(\mathbf{x}) = \sum_{i=1}^{N} x_i^2$$

with $\mathbf{x} \in \mathbb{R}^N$ and the global optimum at $\mathbf{x}^* = \mathbf{0}$ with $f(\mathbf{x}^*) = 0$. This function is used to test the basic convergence properties of an algorithm due to its simple, convex nature.

Appendix B
AI Math

B.1 Vectors and Matrices

Vectors and matrices are fundamental in machine learning for data representation and transformation. A vector is an ordered list of numbers, often written as:

$$\mathbf{v} = \begin{bmatrix} v_1 \\ v_2 \\ \vdots \\ v_n \end{bmatrix}$$

A matrix extends this to two dimensions:

$$\mathbf{A} = \begin{bmatrix} a_{11} & a_{12} & \dots & a_{1n} \\ a_{21} & a_{22} & \dots & a_{2n} \\ \vdots & \vdots & \ddots & \vdots \\ a_{m1} & a_{m2} & \dots & a_{mn} \end{bmatrix}.$$

The dot product of two vectors is:

$$\mathbf{u} \cdot \mathbf{v} = \sum_{i=1}^{n} u_i v_i.$$

The dot product between two vectors is a simple way to measure how similar they are. If two vectors point in the same direction, their dot product is large and positive. If they are orthogonal (unrelated), the dot product is zero. If they point in opposite directions, the dot product is negative.

© The Editor(s) (if applicable) and The Author(s), under exclusive license to Springer Nature Switzerland AG 2026
O. Kramer, *Artificial Intelligence Essentials*,
https://doi.org/10.1007/978-3-032-06637-4

The transpose of a matrix swaps its rows and columns:

$$(\mathbf{A}^\top)_{ij} = (\mathbf{A})_{ji}.$$

The norm of a vector measures its magnitude. The Euclidean norm is:

$$\|\mathbf{v}\| = \sqrt{\sum_{i=1}^{n} v_i^2}$$

Matrix multiplication allows combining linear transformations. Given a matrix $\mathbf{A} \in \mathbb{R}^{m \times n}$ and a matrix $\mathbf{B} \in \mathbb{R}^{n \times p}$, their product $\mathbf{C} = \mathbf{AB} \in \mathbb{R}^{m \times p}$ is defined as:

$$c_{ij} = \sum_{k=1}^{n} a_{ik} b_{kj}$$

For example:

$$\mathbf{A} = \begin{bmatrix} 1 & 2 \\ 3 & 4 \end{bmatrix}, \quad \mathbf{B} = \begin{bmatrix} 5 & 6 \\ 7 & 8 \end{bmatrix}, \quad \mathbf{AB} = \begin{bmatrix} 19 & 22 \\ 43 & 50 \end{bmatrix}$$

B.2 Probability Distributions

A uniform distribution assigns equal probability within a range:

$$f(x) = \begin{cases} \frac{1}{b-a}, & a \le x \le b \\ 0, & \text{otherwise} \end{cases}$$

The Bernoulli distribution models a binary outcome (0 or 1) with success probability p:

$$P(X = x) = \begin{cases} p, & x = 1 \\ 1 - p, & x = 0 \end{cases} \quad \text{for } x \in \{0, 1\}$$

B.3 Chain Rule

The chain rule allows the computation of the derivative of a composite function. For a function $y = f(g(x))$, the derivative of y with respect to x is given by:

$$\frac{\partial y}{\partial x} = \frac{\partial f}{\partial g} \cdot \frac{\partial g}{\partial x}$$

This means that the rate of change of y with respect to x can be found by multiplying the rate of change of f with respect to g by the rate of change of g with respect to x.
Example: Let

$$y = \sin(x^2)$$

Here, $f(u) = \sin(u)$ and $g(x) = x^2$, so:

$$\frac{\partial y}{\partial x} = \cos(x^2) \cdot 2x$$

The chain rule is used to compute gradients through composed functions during backpropagation.

Appendix C
Python Programming

Python is a high-level, interpreted language prized for readability and rapid development. It supports procedural, object-oriented, and functional styles, and has become the go-to language for machine learning, data analysis, and scientific computing thanks to its rich ecosystem.

C.1 Installation and Google Colab

Install Python from https://www.python.org/downloads/ and make sure it is added to your system PATH. Python's package installer `pip` can be used to install essential machine learning libraries. For example:

```
pip install numpy pandas scikit-learn tensorflow
```

This installs `numpy` for numerical arrays, `pandas` for data frames, `scikit-learn` for classical machine learning algorithms, and `tensorflow` for deep learning.

Alternatively, Google Colab (https://colab.research.google.com) provides a free, cloud-based environment for running Python code in Jupyter notebooks. It includes preinstalled libraries and supports GPU/TPU acceleration. Notebooks are stored on Google Drive and can be shared and edited collaboratively.

© The Editor(s) (if applicable) and The Author(s), under exclusive license to Springer   77
Nature Switzerland AG 2026
O. Kramer, *Artificial Intelligence Essentials*,
https://doi.org/10.1007/978-3-032-06637-4

C.2   Data Types

Python provides several fundamental data types that are frequently used in machine learning for representing data, configuration options, and model parameters.

Integers (`int`) represent discrete values such as counts, dimensions, or loop counters.

```
num_epochs = 10
```

Floats (`float`) store continuous values, such as learning rates or loss values.

```
learning_rate = 0.001
```

Strings (`str`) are used for textual data like labels, model names, or file paths.

```
model_name = "RandomForest"
```

Booleans (`bool`) represent binary conditions, often used as flags in control structures or training configurations.

```
shuffle = True
```

Lists (`list`) are ordered sequences of values, commonly used to store feature names, datasets, or hyperparameter options.

```
features = ['age', 'income', 'credit_score']
learning_rates = [0.1, 0.01, 0.001]
```

Lists are zero-indexed and support slicing operations:

```
X_batch = X_train[:100]        # First 100 samples
last_rows = X_train[-10:]      # Last 10 samples
```

They also support list comprehensions, which allow concise transformations:

```
k_values = [1, 3, 5, 7]
squared = [x*x for x in k_values]
```

Tuples (`tuple`) are immutable, fixed-length collections, often used to group related values like split ratios.

```
split = (0.8, 0.2)
```

Dictionaries (`dict`) are key-value mappings, used to store model parameters or encode categorical labels.

```
params = {'n_estimators': 100, 'max_depth': 5}
label_map = {'cat': 0, 'dog': 1}
```

C.3   Loops and Conditionals

Loops are used to repeat operations over sequences or ranges. In machine learning, they are essential for training models over multiple epochs and batches.

```python
# Training loop over epochs and batches
for epoch in range(num_epochs):
    for batch in data_loader:
        train_on(batch)
```

Conditional statements control program flow based on logical decisions. They are commonly used for model evaluation and early stopping.

```python
if val_loss < best_loss:
    best_loss = val_loss
    save_model()
elif epoch - last_improve > patience:
    break
```

C.4   From Functions to Classes

Functions define reusable blocks of logic. In machine learning, they help organize evaluation, preprocessing, or metric computations.

```python
def evaluate(model, X, y):
    preds = model.predict(X)
    return accuracy_score(y, preds)
```

Functional programming provides compact syntax for transforming data. The map function applies an operation to each element, while filter selects elements based on a condition.

```python
# Normalize features
normalized = list(map(lambda x: x / max_val, feature_list))
# Filter even values
positive = list(filter(lambda x: x % 2 == 0, scores))
```

Object-oriented design allows bundling model logic into classes. This is useful for encapsulating training, prediction, and parameters.

```python
class Classifier:
    def __init__(self, **params):
        self.model = RandomForestClassifier(**params)
    def train(self, X, y):
        self.model.fit(X, y)
    def predict(self, X):
        return self.model.predict(X)
```

C.5   Essential Python Tools

Modern machine learning workflows rely on a variety of Python libraries that simplify data handling, model development, and experimentation. This section introduces several core tools, including NumPy, scikit-learn, Keras, and Conda environments.

NumPy provides efficient array structures and vectorized operations. It serves as the numerical foundation for most ML libraries.

```python
import numpy as np
X = np.array([[1,2,3], [4,5,6]])
means = X.mean(axis=0)
```

Scikit-learn offers a high-level interface for classical machine learning tasks, including classification, regression, and clustering. It provides consistent APIs for model training, prediction, and evaluation.

```python
from sklearn.ensemble import RandomForestClassifier
model = RandomForestClassifier(n_estimators=100)
model.fit(X_train, y_train)
preds = model.predict(X_test)
```

Keras, built on top of TensorFlow, allows fast prototyping and training of deep learning models. A typical neural network is created by stacking layers and compiling the model with a loss function and optimizer.

```python
from tensorflow.keras import Sequential
from tensorflow.keras.layers import Dense
model = Sequential([
    Dense(64, activation='relu', input_shape=(784,)),
    Dense(10, activation='softmax')
])
model.compile(optimizer='adam',
              loss='sparse_categorical_crossentropy')
model.fit(X_train, y_train, epochs=5)
```

To keep dependencies organized and experiments reproducible, it is best practice to work in isolated environments. Conda is a popular tool for managing such environments, especially when working with different versions of Python and third-party libraries.

```
conda create -n ml python=3.10 numpy pandas scikit-learn
conda activate ml
```

Together, these tools provide a powerful ecosystem for building, training, and deploying machine learning models.

References

1. Adams D (1979) The Hitchhiker's guide to the galaxy. Pan Books, London. ISBN: 9780330258647
2. Ba JL, Kiros JR, Hinton GE (2016) Layer normalization. arXiv: 1607.06450 [stat.ML]
3. Beyer H-G, Schwefel H-P (2002) Evolution strategies- a comprehensive introduction. Natural Comput 1:3–52
4. Bishop CM (2006) Pattern recognition and machine learning. Springer
5. Christiano PF et al (2017) Deep reinforcement learning from human preferences. In: Advances in neural information processing systems (NeurIPS), pp 4299–4307
6. Cobbe K et al (2021) Training verifiers to solve math word problems. https://arxiv.org/abs/2110.14168
7. Cover TM, Hart PE (1967) Nearest neighbor pattern classification. IEEE Tran Inf Theory 13(1):21–27. https://doi.org/10.1109/TIT.1967.1053964
8. Devlin J et al (2019) BERT: pre-training of deep bidirectional trans formers for language understanding. In: Proceedings of the 2019 Conference of the North American chapter of the association for computational linguistics: human language technologies, vol. 1. Association for Computational Linguistics, pp 4171–4186
9. Goertzel B (2014) Artificial general intelligence: concept, state of the art, and future prospects. J Artif General Intell 5(1):1–48. https://doi.org/10.2478/jagi-2014-0001
10. Goodfellow I, Bengio Y, Courville A (2016) Deep learning. MIT Press
11. Hastie T, Tibshirani R, Friedman J (2009) The elements of statistical learning. Springer. https://doi.org/10.1007/978-0-387-84858-7
12. He K et al (2016) Deep residual learning for image recognition. In: Proceedings of the IEEE conference on computer vision and pattern recognition (CVPR). IEEE, pp 770–778
13. Hendrycks D et al (2021) Measuring mathematical problem solving with the MATHDataset. In: Van schoren J, Yeung S-K (ed) Proceedings of the neural information processing systems track on datasets and benchmarks 1, NeurIPS Datasets and Benchmarks 2021, December 2021, virtual. https://datasets-benchmarks-proceedings.neurips.cc/paper/2021/hash/
14. Holland JH (1992) Adaptation in natural and artificial systems: an introductory analysis with applications to biology, control, and artificial intelligence. MIT Press, Cambridge, MA. https://doi.org/10.7551/mitpress/1090.001.0001
15. James G et al (2013). An introduction to statistical learning. Springer. https://doi.org/10.1007/978-1-4614-7138-7
16. Kramer O (2017) Genetic algorithm essentials, vol 679. Studies in computational intelligence. Springer. ISBN: 978-3-319-52155-8. https://doi.org/10.1007/978-3-319-52156-5

© The Editor(s) (if applicable) and The Author(s), under exclusive license to Springer Nature Switzerland AG 2026
O. Kramer, *Artificial Intelligence Essentials*,
https://doi.org/10.1007/978-3-032-06637-4

17. Krizhevsky A, Sutskever I, Hinton GE (2012) ImageNet classification with deep convolutional neural networks. In: Advances in neural information processing systems (NeurIPS), pp 1106–1114

18. LeCun Y et al (2012) Efficient BackProp. In: Montavon G, Orr GB, Müller K-R (ed) Neural networks: tricks of the trade, 2nd edn, vol 7700. Lecture notes in computer science. Springer, pp 9–48. https://doi.org/10.1007/978-3-642-35289-8_3

19. Cunetal YL (1998) Gradient-based learning applied to document recognition. Proc IEEE 86(11):2278–2324. https://doi.org/10.1109/5.726791

20. Lewis P et al (2020) Retrieval-augmented generation for knowledge intensive NLP tasks. In: Advances in neural information processing systems (NeurIPS)

21. Lloyd SP (1982) Least squares quantization in PCM. IEEE Trans Inf Theory 28(2):129–137. https://doi.org/10.1109/TIT.1982.1056489

22. Mialon G et al (2023) GAIA: a benchmark for general AI assistants. arXiv: 2311.12983

23. Pedregosa F et al (2011) Scikit-learn: machine learning in Python. J Mach Learn Res 12:2825–2830. https://doi.org/10.5555/1953048.2078195

24. Radford A et al (2018) Improving language understanding by generative pre-training. OpenAI, pp 1–12

25. Radford A et al (2023) Robust speech recognition via large-scale weak supervision. In: International conference on machine learning (ICML), vol 202. Proceedings of machine learning research, pp 28492–28518

26. Rosenblatt F (1958) The perceptron: a probabilistic model for information storage and organization in the Brain. Psychol. Rev. 65(6):386–408

27. Rumelhart DE, Hinton GE, Williams RJ (1986) Learning representations by back-propagating errors. Nature 323(6088):533–536

28. Russell S, Norvig P (1995) Artificial intelligence: a modern approach. Pearson

29. Schwefel H-P (1995) Evolution and optimum seeking. Wiley, New York. ISBN: 9780471940806

30. Shorten C, Khoshgoftaar TM (2019) A survey on image data augmentation for deep learning. J Big Data 6(1):1–48

31. Simonyan K, Zisserman A (2015) Very deep convolutional networks for large-scale image recognition. In: International conference on learning representations (ICLR). http://arxiv.org/abs/1409.1556

32. Srivastava N et al (2014) Dropout: a simple way to prevent neural networks from overfitting. J Mach Lear Res 15(1):1929–1958

33. Szegedy C et al (2015) Going deeper with convolutions. In: IEEE conference on computer vision and pattern recognition (CVPR). IEEE Computer Society, pp 1–9

34. Touvron H et al (2023) LLaMA: open and efficient foundation language models. arXiv:2302.13971. https://doi.org/10.48550/arXiv.2302.13971

35. Turing AM (1950) Computing machinery and intelligence. Mind 59(236):433–460

36. Vaswani A et al (2017) Attention is all you need. In: Advances in neural information processing systems (NeurIPS), pp 5998–6008. https://proceedings.neurips.cc/paper/2017/hash/3f5ee243547dee91fbd053c1c4a845aa-Abstract.html

37. Wei J et al (2022) In: Advances in neural information processing systems (NeurIPS). http://papers.nips.cc/paper_files/paper/2022/hash/9d5609613524ecf4f15af0f7b31abca4-Abstract-Conference.html

38. Zhang A et al (2020) Dive into deep learning. https://d2l.ai.self published

Batch number: 10146286

Printed by Printforce, the Netherlands